Climate Change Indicators in the United States, 2014

Third Edition

TECHNICAL DOCUMENTATION

May 2014

Contents

Overview

This document provides technical supporting information for the 30 indicators and four chapter-specific call-out features that appear in the U.S. Environmental Protection Agency's (EPA's) report, *Climate Change Indicators in the United States, 2014*. EPA prepared this document to ensure that each indicator is fully transparent—so readers can learn where the data come from, how each indicator was calculated, and how accurately each indicator represents the intended environmental condition. EPA uses a standard documentation form, then works with data providers and reviews the relevant literature and available documentation associated with each indicator to address the elements on the form as completely as possible.

EPA's documentation form addresses 13 elements for each indicator:

1. Indicator description
2. Revision history
3. Data sources
4. Data availability
5. Data collection (methods)
6. Indicator derivation (calculation steps)
7. Quality assurance and quality control (QA/QC)
8. Comparability over time and space
9. Data limitations
10. Sources of uncertainty (and quantitative estimates, if available)
11. Sources of variability (and quantitative estimates, if available)
12. Statistical/trend analysis (if any has been conducted)
13. References

In addition to indicator-specific documentation, this appendix to the report summarizes the criteria that EPA uses to screen and select indicators for inclusion in reports. This documentation also describes the process EPA follows to select and develop those indicators that have been added or substantially revised since the publication of EPA's first version of this report in April 2010. Indicators that are included in the report must meet all of the criteria. Lastly, this document provides general information on changes that have occurred since the 2012 version of the *Climate Indicators in the United States* report.

The development of the indicators report, including technical documentation, was conducted in accordance with EPA's *Guidelines for Ensuring and Maximizing the Quality, Objectivity, Utility, and Integrity of Information Disseminated by the Environmental Protection Agency.*[1]

EPA may update this technical documentation as new and/or additional information about these indicators and their underlying data becomes available. Please contact EPA at: climateindicators@epa.gov to provide any comments about this documentation.

---

[1] U.S. EPA. 2002. Guidelines for ensuring and maximizing the quality, objectivity, utility, and integrity of information disseminated by the Environmental Protection Agency. EPA/260R-02-008. http://www.epa.gov/quality/informationguidelines/documents/EPA_InfoQualityGuidelines.pdf.

EPA's Indicator Evaluation Criteria

General Assessment Factors

When evaluating the quality, objectivity, and relevance of scientific and technical information, the considerations that EPA takes into account can be characterized by five general assessment factors, as found in *A Summary of General Assessment Factors for Evaluating the Quality of Scientific and Technical Information*.[2] These general assessment factors and how EPA considers them in development of climate change indicators are:

- **Soundness (AF1)** is defined as the extent to which the scientific and technical procedures, measures, methods, or models employed to generate the information are reasonable for and consistent with the intended application. As described below, EPA follows a process that carefully considers 10 criteria for each proposed indicator. EPA evaluates the scientific and technical procedures, measures, and methods employed to generate the data that underpin each indicator as part of its consideration of the 10 selection criteria. If a proposed indicator and associated data meet all of the criteria, EPA determines they are reasonable for, and consistent with, use as an indicator for this report.

- **Applicability and utility (AF2)** is defined as the extent to which the information is relevant for the Agency's intended use. Considerations related to this assessment factor include the relevance of the indicator's purpose, design, outcome measures, results, and conditions to the Agency's intended use. As described below, EPA follows a process that carefully considers 10 criteria for each proposed indicator. Some of these criteria relate to the relevance or usefulness of the indicator.

- **Clarity and completeness (AF3)** is defined as the degree of clarity and completeness with which the data, assumptions, methods, quality assurance, sponsoring organizations, and analyses employed to generate the information are documented. EPA investigates each indicator's underlying data, assumptions, methods and analyses employed to generate the information, quality assurance, and sponsoring organizations in order to record this information clearly, completely, and transparently in a publicly available technical support document. Because the underlying data and methods for analyses are peer-reviewed and/or published by federal agencies and reputable scientific journals, these publications provide additional documentation of assumptions, methods, and analyses employed to generate the information.

- **Uncertainty and variability (AF4)** is defined as the extent to which the variability and uncertainty (quantitative and qualitative) in the information or in the procedures, measures, methods, or models are evaluated and characterized. EPA carefully considers the extent to which the uncertainty and variability of each indicator's underlying data were evaluated and characterized, based on their underlying documentation and source publications. EPA also

---

[2] U.S. EPA. 2003. Science Policy Council assessment factors: A summary of general assessment factors for evaluating the quality of scientific and technical information. EPA 100/B-03/001. www.epa.gov/stpc/pdfs/assess2.pdf.

describes all sources of uncertainty and variability, as well as data limitations (see elements #9, #10, and #11, listed above) in the technical documentation.

- **Evaluation and review (AF5)** is defined as the extent of independent verification, validation, and peer review of the information or of the procedures, measures, methods, or models. EPA carefully considers the extent to which the data underlying each indicator are independently verified, validated, and peer-reviewed. One of EPA's selection criteria relates to peer review of the data and methods associated with the indicator. EPA also ensures that each edition of the report—including supporting technical documentation—is independently peer-reviewed.

The report and associated technical documentation are consistent with guidance discussed in a newer document, *Guidance for Evaluating and Documenting the Quality of Existing Scientific and Technical Information*,[3] issued in December 2012 as an addendum to the 2003 EPA guidance document. These general assessment factors form the basis for the 10 criteria EPA uses to evaluate indicators, which are documented in 13 elements as part of the technical documentation. These 13 elements are mapped to EPA's criteria and the assessment factors in the table below.

Criteria for Including Indicators in This Report

EPA used a set of 10 criteria to carefully select indicators for inclusion in the *Climate Change Indicators in the United States, 2014* report. The following table introduces these criteria and describes how they relate to the five general assessment factors and the 13 elements in EPA's indicator documentation form, both listed above.

---

[3] U.S. EPA. 2012. Guidance for evaluating and documenting the quality of existing scientific and technical information. www.epa.gov/stpc/pdfs/assess3.pdf.

Assessment Factor	Criterion	Description	Documentation Elements
AF1, AF2, AF4	**Trends over time**	Data are available to show trends over time. Ideally, these data will be long-term, covering enough years to support climatically relevant conclusions. Data collection must be comparable across time and space. Indicator trends have appropriate resolution for the data type.	4. Data availability 5. Data collection 6. Indicator derivation
AF1, AF2, AF4	**Actual observations**	The data consist of actual measurements (observations) or derivations thereof. These measurements are representative of the target population.	5. Data collection 6. Indicator derivation 8. Comparability over time and space 12. Statistical/ trend analysis
AF1, AF2	**Broad geographic coverage**	Indicator data are national in scale or have national significance. The spatial scale is adequately supported with data that are representative of the region/area.	4. Data availability 5. Data collection 6. Indicator derivation 8. Comparability over time and space
AF1, AF3, AF5	**Peer-reviewed data** (peer-review status of indicator and quality of underlying source data)	Indicator and underlying data are sound. The data are credible, reliable, and have been peer-reviewed and published.	3. Data sources 4. Data availability 5. Data collection 6. Indicator derivation 7. QA/QC 12. Statistical/ trend analysis
AF4	**Uncertainty**	Information on sources of uncertainty is available. Variability and limitations of the indicator are understood and have been evaluated.	5. Data collection 6. Indicator derivation 7. QA/QC 9. Data limitations 10. Sources of uncertainty 11. Sources of variability 12. Statistical/ trend analysis
AF1, AF2	**Usefulness**	Indicator informs issues of national importance and addresses issues important to human or natural systems. Complements existing indicators.	6. Indicator derivation

Assessment Factor	Criterion	Description	Documentation Elements
AF1, AF2	**Connection to climate change**	The relationship between the indicator and climate change is supported by published, peer-reviewed science and data. A climate signal is evident among stressors, even if the indicator itself does not yet show a climate signal. The relationship to climate change is easily explained.	6. Indicator derivation 11. Sources of variability
AF1, AF3, AF4, AF5	**Transparent, reproducible, and objective**	The data and analysis are scientifically objective and methods are transparent. Biases, if known, are documented, minimal, or judged to be reasonable.	4. Data availability 5. Data collection 6. Indicator derivation 7. QA/QC 9. Data limitations 10. Sources of uncertainty 11. Sources of variability
AF2, AF3	**Understandable to the public**	The data provide a straightforward depiction of observations and are understandable to the average reader.	6. Indicator derivation 9. Data limitations
AF2	**Feasible to construct**	The indicator can be constructed or reproduced within the timeframe for developing the report. Data sources allow routine updates of the indicator for future reports.	3. Data sources 4. Data availability 5. Data collection 6. Indicator derivation

Process for Evaluating Indicators for the 2012 and 2014 Reports

EPA published the first edition of *Climate Change Indicators in the United States* in April 2010, featuring 24 indicators. In 2012, EPA published a second edition, using the following approach to identify and develop a robust set of new and revised indicators for the report:

A. Identify and develop a list of candidate indicators.
B. Conduct initial research; screen against a subset of indicator criteria.
C. Conduct detailed research; screen against the full set of indicator criteria.
D. Select indicators for development.
E. Develop draft indicators.
F. Facilitate expert review of draft indicators.
G. Periodically re-evaluate indicators.

EPA followed this same approach to develop additional indicators and a set of features that highlight specific regions or areas of interest for the 2014 report. See Section E below for a description of these features.

In selecting and developing the climate change indicators included in this report, EPA fully complied with the requirements of the Information Quality Act (also referred to as the Data Quality Act) and EPA's *Guidelines for Ensuring and Maximizing the Quality, Objectivity, Utility, and Integrity of Information Disseminated by the Environmental Protection Agency.*[4]

The process for evaluating indicators is described in more detail below.

A: Identify Candidate Indicators

EPA investigates and vets new candidate indicators through coordinated outreach, stakeholder engagement, and reviewing the latest scientific literature. New indicators and content can be broadly grouped into two categories:

- Additions: Completely new indicators.
- Revisions: Improving an existing indicator by adding or replacing metrics or underlying data sources. These revisions involve obtaining new data sets and vetting their scientific validity.

Outreach and Stakeholder Engagement

EPA invited suggestions of new indicators from the public following the release of the April 2010 *Climate Change Indicators in the United States* report, and continues to welcome suggestions at

---

[4] U.S. EPA. 2002. Guidelines for ensuring and maximizing the quality, objectivity, utility, and integrity of information disseminated by the Environmental Protection Agency. EPA/260R-02-008. www.epa.gov/quality/informationguidelines/documents/EPA_InfoQualityGuidelines.pdf.

climateindicators@epa.gov. In March 2011, EPA held an information gathering meeting of experts on climate change and scientific communication to obtain their feedback on the first edition of the report. Meeting participants considered the merits of data in the report and provided suggestions for new and revised content. Participants suggested a variety of concepts for new indicators and data sources for EPA to consider.

Suggestions from the participants informed EPA's investigation into candidate indicators for the screening and selection process. As part of this process, existing indicators are re-evaluated as appropriate to ensure that they continue to function as intended and meet EPA's indicator criteria.

New Science and Data

The process of identifying indicators includes monitoring the scientific literature, assessing the availability of new data, and eliciting expert review. Many federal agencies and other organizations have ongoing efforts to make new data available, which allows for continued investigation into opportunities for compiling or revising indicator content. EPA also engages with existing data contributors and partners to help improve existing indicators and identify potential new indicators.

B and C: Research and Screening

Indicator Criteria

EPA screens and selects indicators based on an objective, transparent process that considers the scientific integrity of each candidate indicator, the availability of data, and the value of including the candidate indicator in the report. Each candidate indicator is evaluated against fundamental criteria to assess whether or not it is reasonable to further evaluate and screen the indicator for inclusion in the upcoming report. These fundamental criteria include: peer-review status of the data, accessibility of the underlying data, relevance and usefulness of the indicator (i.e., the indicator's ability to be understood by the public), and its connection to climate change.

Tier 1 Criteria

- Peer-reviewed data
- Feasible to construct
- Usefulness
- Understandable to the public
- Connection to climate change

Tier 2 Criteria

- Transparent, reproducible, and objective
- Broad geographic range
- Actual observations
- Trends over time
- Uncertainty

The distinction between Tier 1 and Tier 2 criteria is not intended to suggest that one group is necessarily more important than the other. Rather, EPA determined that a reasonable approach was to consider which criteria must be met before proceeding further and to narrow the list of indicator candidates before the remaining criteria were applied.

Screening Process

EPA researches and screens candidate indicators by creating and populating a database comprising all suggested additions and revisions, then documents the extent to which each of these candidate indicators meet each of EPA's criteria. EPA conducts the screening process in two stages:

- **Tier 1 screening:** Indicators are evaluated against the set of Tier 1 criteria. Indicators that reasonably meet these criteria are researched further; indicators that do not meet these criteria are eliminated from consideration. Some of the candidate indicators ruled out at this stage are ideas that could be viable indicators in the future (e.g., indicators that do not yet have published data or need further investigation into methods).

- **Tier 2 screening:** Indicators deemed appropriate for additional screening are assessed against the Tier 2 criteria. Based on the findings from the complete set of 10 criteria, the indicators are again evaluated based on the assessment of the remaining criteria.

Information Sources

To assess each candidate indicator against the criteria, EPA reviews the scientific literature using numerous methods (including several online databases and search tools) to identify existing data sources and peer-reviewed publications.

In cases where the candidate indicator is not associated with a well-defined metric, EPA conducts a broader survey of the literature to identify the most frequently used metrics. For instance, an indicator related to "community composition" (i.e., biodiversity) was suggested, but it was unclear how this variable might best be measured or represented by a metric.

As noted above, to gather additional information, EPA contacts appropriate subject matter experts, including authors of identified source material, existing data contributors, and collaborators.

D: Indicator Selection

Based on the results of the screening process, the most promising indicators for the report are developed into proposed indicator summaries. EPA consults the published literature, subject matter experts, and online databases to obtain data for each of these indicators. Upon acquiring sound data and technical documentation, EPA prepares a set of possible graphics for each indicator, along with a summary table that describes the proposed metric(s), data sources, limitations, and other relevant information.

Summary information is reviewed by EPA technical staff, and then the indicator concepts that meet the screening criteria are formally approved for development and inclusion in the report.

E: Indicator Development

Approved new and revised indicators are then developed within the framework of the indicator report. Graphics, summary text, and technical documentation for all of the proposed new or revised indicators are developed in accordance with the format established for the original 24 indicators in the 2010 indicators report. An additional priority for development is to make sure that each indicator communicates effectively to a non-technical audience without misrepresenting the underlying data and source(s) of information.

EPA's 2014 report contains "Community Connection" and "A Closer Look" features in certain chapters (e.g., Cherry Blossom Bloom Dates in Washington, D.C.), which focus on a particular region or localized area of interest to augment the current report and engage readers in particular areas of the United States. While the features and their underlying data are not national in scale or representative of broad geographic areas, these features were screened, developed, and documented in a manner consistent with the indicators in the report.

F: Internal and External Reviews

The complete indicator packages (graphics, summary text, and technical documentation) undergo internal review, data provider/collaborator review, and an independent peer review.

Internal Review

The report contents are reviewed at various stages of development in accordance with EPA's standard review protocols for publications. This process includes review by EPA technical staff and various levels of management within the Agency.

Data Provider/Collaborator Review

Organizations and individuals who collected and/or compiled the data (e.g., the National Oceanic and Atmospheric Administration and the U.S. Geological Survey) also review the report.

Independent Peer Review

The peer review of the report and technical supporting information followed the procedures in *EPA's Peer Review Handbook,* 3rd Edition (EPA/100/B-06/002)[5] for reports that do not provide influential scientific information. The review was managed by a contractor under the direction of a designated EPA peer review leader, who prepared a peer review plan, the scope of work for the review contract, and the charge for the reviewers. The peer review leader played no role in producing the draft report, except for writing the technical support documentation for the feature, Land Loss Along the Atlantic Coast, and he recused himself from all matters relating to that feature.

Under the general approach of the peer review plan, the peer review consisted of 12 experts:

---

[5] U.S. EPA. 2006. EPA's peer review handbook. Third edition. EPA 100/B-06/002. www.epa.gov/peerreview/pdfs/peer_review_handbook_2006.pdf.

- One expert with specific expertise in environmental indicators reviewed the entire report.
- One expert with specific expertise in environmental communication reviewed the entire report.
- One general expert in the causes and effects of climate change reviewed the entire report.
- Nine subject matter experts each reviewed a single chapter or selected indicators within their fields of expertise. Those experts had the following expertise: greenhouse gas emissions, atmospheric ozone, climate science and meteorology, ocean acidity, coastal land cover data, snow cover and snowfall, wildfires, bird wintering ranges, and Lyme disease.

The peer review charge asked reviewers to provide detailed comments and to indicate whether the report (or chapter) should be published (a) as-is, (b) with changes suggested by the review, (c) only after a substantial revision necessitating a re-review, or (d) not at all. Ten reviewers answered (a) or (b), while two reviewers answered (c). A third reviewer with expertise in meteorology was satisfied with the indicators she reviewed, but suggested making clarifications in the introductory chapter and the introduction to every indicator.

Although the three reviewers' fundamental concerns differed, they shared a common theme, which was that the climate change indicators in EPA's report should convey a climate signal, which the report should explain. The reviewers of the Lyme disease and wildfire indicators expressed concerns about whether the data series used by those two indicators had been constructed to demonstrate a climate signal as effectively as alternative formulations that they suggested. The meteorology reviewer suggested that the report should distinguish between climate change in general and anthropogenic climate change in particular, to avoid confusion between detection of a change and attribution to human causes.

EPA revised the report to address all comments and prepared a spreadsheet to document the response to each of the approximately 600 comments from the peer review. The revised report and EPA's responses were then sent for re-review to the three reviewers who had expressed reservations during the initial review, and the general climate change expert was asked to review EPA's responses to those three reviews.

The general climate change expert concluded that EPA had adequately addressed the comments from the three indicator-specific reviewers who had sought re-review. The wildfire reviewer concluded that the revised indicator was accurate, but he remained concerned that because it is based on a nationwide tally of wildfires, EPA's indicator cannot effectively detect climate signals, while a more regionally explicit indicator might provide more information on this issue. The meteorology reviewer thought the report needed further clarification of whether the indicators reflect anthropogenic climate change or all (natural and anthropogenic) climate change. The Lyme disease reviewer concluded that the indicator still required fundamental revision before it could distinguish a possible signal of climate change from other causes of the changing incidence of Lyme disease.

These comments led EPA to make additional revisions to address the theme that ran through all three critical reviews: EPA had never intended to suggest that all climate change indicators include a detectable signal of the effects of climate change. Indicators are sometimes monitored because they are expected to *eventually* show an effect, not because they already do so. The draft report had not thoroughly addressed this distinction, so it was possible to read the draft report as implying that climate change was the primary causal factor for each indicator. EPA revised the report's introduction and the technical support documentation for the Lyme disease and wildfire indicators to emphasize that the indicators do not necessarily reflect a detectable impact of climate change on the system represented by a given indicator.

EPA's peer-review leader conducted a quality control check to ensure that the authors took sufficient action and provided an adequate response for every peer-review and re-review comment. Based on the findings, the authors addressed those comments with changes to the report and/or changes to the response to those comments.

G: Periodic Re-Evaluation of Indicators

Existing indicators are evaluated to ensure they are relevant, comprehensive, and sustainable. The process of evaluating indicators includes monitoring the availability of newer data, eliciting expert review, and assessing indicators in light of new science. For example, EPA determined that the underlying methods for developing the Plant Hardiness Zone indicator that appeared in the first edition of *Climate Change Indicators in the United States* (April 2010) had significantly changed, such that updates to the indicator are no longer possible. Thus, EPA removed this indicator from the 2012 edition. EPA re-evaluates indicators during the time between publication of the reports.

EPA updated several existing indicators with additional years of data, new metrics or data series, and analyses based on data or information that have become available since the publication of EPA's 2012 report. For example, EPA was able to update the Global Greenhouse Gas Emissions indicator with land use, land-use change, and forestry data that have now become available through the data source used for this indicator. These and other revisions are documented in the technical documentation specific to each indicator.

Summary of Changes to the 2014 Report

The table below highlights major changes made to the indicators during development of the 2014 version of the report, compared with the 2012 report.

Indicator (number of figures)	Change	Years of data added since 2012 report	Most recent data
U.S. Greenhouse Gas Emissions (3)		2	2012
Global Greenhouse Gas Emissions (3)	Added land use and forestry	3	2011
Atmospheric Concentrations of Greenhouse Gases (5)	Expanded with new metric (ozone); extended to 800,000 BC	2	2013
Climate Forcing (2)	Expanded with new metric (broader set of human activities)	2	2013
U.S. and Global Temperature (3)		2	2013
High and Low Temperatures (6)	Expanded with new metrics (95th and 5th percentiles)	2	2014
U.S. and Global Precipitation (3)		1	2012
Heavy Precipitation (2)		2	2013
Drought (2)		2	2013
Tropical Cyclone Activity (3)		2	2013
Ocean Heat (1)		2	2013
Sea Surface Temperature (2)	Replaced example map with new metric (map of change over time)	2	2013
Sea Level (2)		2	2013
Ocean Acidity (2)		1	2013
Arctic Sea Ice (2)		1	2013
Glaciers (2)		2	2012
Lake Ice (3)	Replaced duration graph with thaw date trend map	2	2012
Snowfall (2)		3	2014
Snow Cover (2)		2	2013
Snowpack (1)		13	2013
Heating and Cooling Degree Days (3)	New indicator		2013
Heat-Related Deaths (1)		1	2010
Lyme Disease (2)	New indicator		2012
Length of Growing Season (3)		2	2013
Ragweed Pollen Season (1)		2	2013
Wildfires (4)	New indicator		2013
Streamflow (4)	Expanded with new metric (annual average streamflow)	3	2012
Great Lakes Water Levels and Temperatures (2)	New indicator		2013
Bird Wintering Ranges (2)		8	2013
Leaf and Bloom Dates (3)	Combined graphs; added two trend maps	3	2013

Discontinued Indicators

Plant Hardiness Zones: Discontinued in April 2012

Reason for Discontinuation:

This indicator compared the U.S. Department of Agriculture's (USDA's) 1990 Plant Hardiness Zone Map (PHZM) with a 2006 PHZM that the Arbor Day Foundation compiled using similar methods. USDA developed[6] and published a new PHZM in January 2012, reflecting more recent data as well as the use of better analytical methods to delineate zones between weather stations, particularly in areas with complex topography (e.g., many parts of the West). Because of the differences in methods, it is not appropriate to compare the original 1990 PHZM with the new 2012 PHZM to assess change, as many of the apparent zone shifts would reflect improved methods rather than actual temperature change. Further, USDA cautioned users against comparing the 1990 and 2012 PHZMs and attempting to draw any conclusions about climate change from the apparent differences.

For these reasons, EPA chose to discontinue the indicator. EPA will revisit this indicator in the future if USDA releases new editions of the PHZM that allow users to examine changes over time.

For more information about USDA's 2012 PHZM, see: http://planthardiness.ars.usda.gov/PHZMWeb/. The original version of this indicator as it appeared in EPA's 2010 report can be found at: www.epa.gov/climatechange/indicators/download.html.

---

[6] Daly, C., M.P. Widrlechner, M.D. Halbleib, J.I. Smith, and W.P. Gibson. 2012. Development of a new USDA plant hardiness zone map for the United States. J. Appl. Meteorol. Clim. 51:242–264.

U.S. Greenhouse Gas Emissions

Identification

1. Indicator Description

This indicator describes emissions of greenhouse gases (GHGs) in the United States and its territories between 1990 and 2012. This indicator reports emissions of GHGs according to their global warming potential, a measure of how much a given amount of the GHG is estimated to contribute to global warming over a selected period of time. For the purposes of comparison, global warming potential values are given in relation to carbon dioxide (CO_2) and are expressed in terms of CO_2 equivalents. This indicator is highly relevant to climate change because greenhouse gases from human activities are the primary driver of observed climate change since the mid-20th century (IPCC, 2013).

Components of this indicator include:

- U.S. GHG emissions by gas (Figure 1).
- U.S. GHG emissions and sinks by economic sector (Figure 2).
- U.S. GHG emissions per capita and per dollar of GDP (Figure 3).

2. Revision History

April 2010:   Indicator posted.
December 2011:   Updated with data through 2009.
April 2012:   Updated with data through 2010.
August 2013:   Updated on EPA's website with data through 2011.
April 2014:   Updated with data through 2012.

Data Sources

3. Data Sources

This indicator uses data and analysis from EPA's *Inventory of U.S. Greenhouse Gas Emissions and Sinks* (U.S. EPA, 2014), an assessment of the anthropogenic sources and sinks of GHGs for the United States and its territories for the period from 1990 to 2012.

4. Data Availability

The complete U.S. GHG inventory is published annually, and the version used to prepare this indicator is publicly available at: www.epa.gov/climatechange/ghgemissions/usinventoryreport.html (U.S. EPA, 2014). The figures in this indicator are taken from the following figures and tables in the inventory report:

- Figure 1 (emissions by gas): Figure ES-1/Table ES-2.
- Figure 2 (emissions by economic sector): Figure ES-13/Table ES-7.

- Figure 3 (emissions per capita and per dollar gross domestic product [GDP]): Figure ES-15/Table ES-9.

The inventory report does not present data for the years 1991–2004 or 2006–2007 due to space constraints. However, data for these years can be obtained by downloading the complete supporting tables or by contacting EPA's Climate Change Division (www.epa.gov/climatechange/contactus.html).

Figure 3 includes trends in population and real GDP. EPA obtained publicly available population data from the U.S. Census Bureau's International Data Base at: www.census.gov/population/international/. EPA obtained GDP data from the U.S. Department of Commerce, Bureau of Economic Analysis. These data are publicly available from the Bureau of Economic Analysis website at: www.bea.gov/national/index.htm#gdp.

Methodology

5. Data Collection

This indicator uses data directly from the *Inventory of U.S. Greenhouse Gas Emissions and Sinks* (U.S. EPA, 2014). The inventory presents estimates of emissions derived from direct measurements, aggregated national statistics, and validated models. Specifically, this indicator focuses on the six long-lived greenhouse gases currently covered by agreements under the United Nations Framework Convention on Climate Change (UNFCCC). These compounds are CO_2, methane (CH_4), nitrous oxide (N_2O), selected hydrofluorocarbons (HFCs), selected perfluorocarbons (PFCs), and sulfur hexafluoride (SF_6).

The emissions and source activity data used to derive the emissions estimates are described thoroughly in EPA's inventory report. The scientifically approved methods can be found in the Intergovernmental Panel on Climate Change's (IPCC's) GHG inventory guidelines (http://www.ipcc-nggip.iges.or.jp/public/2006gl/index.html) (IPCC, 2006) and in IPCC's *Good Practice Guidance and Uncertainty Management in National Greenhouse Gas Inventories* (www.ipcc-nggip.iges.or.jp/public/gp/english) (IPCC, 2000). More discussion of the sampling and data sources associated with the inventory can be found at: www.epa.gov/climatechange/ghgemissions.

The U.S. GHG inventory provides a thorough assessment of the anthropogenic emissions by sources and removals by sinks of GHGs for the United States from 1990 to 2012. Although the inventory is intended to be comprehensive, certain identified sources and sinks have been excluded from the estimates (e.g., CO_2 from burning in coal deposits and waste piles, CO_2 from natural gas processing). Sources are excluded from the inventory for various reasons, including data limitations or an incomplete understanding of the emissions process. The United States is continually working to improve understanding of such sources and seeking to find the data required to estimate related emissions. As such improvements are made, new emissions sources are quantified and included in the inventory. For a complete list of excluded sources, see Annex 5 of the U.S. GHG inventory report (www.epa.gov/climatechange/ghgemissions/usinventoryreport.html).

Figure 3 of this indicator compares emissions trends with trends in population and U.S. GDP. Population data were collected by the U.S. Census Bureau. For this indicator, EPA used midyear estimates of the total U.S. population. GDP data were collected by the U.S. Department of Commerce, Bureau of

Economic Analysis. For this indicator, EPA used real GDP in chained 2005 dollars, which means the numbers have been adjusted for inflation. See: www.census.gov/population/international for the methods used to determine midyear population estimates for the United States. See: www.bea.gov/methodologies/index.htm#national_meth for the methods used to determine GDP.

6. Indicator Derivation

The U.S. GHG inventory was constructed following scientific methods described in the Intergovernmental Panel on Climate Change's (IPCC's) *Guidelines for National Greenhouse Gas Inventories* (IPCC, 2006) and in IPCC's *Good Practice Guidance and Uncertainty Management in National Greenhouse Gas Inventories* (IPCC, 2000). EPA's annual inventory reports and IPCC's inventory development guidelines have been extensively peer reviewed and are widely viewed as providing scientifically sound representations of GHG emissions.

U.S. EPA (2014) provides a complete description of methods and data sources that allowed EPA to calculate GHG emissions for the various industrial sectors and source categories. Further information on the inventory design can be obtained by contacting EPA's Climate Change Division (www.epa.gov/climatechange/contactus.html).

The inventory covers U.S. GHG data for the years 1990 to 2012, and no attempt has been made to incorporate other locations or to project data forward or backward from this time window. Some extrapolation and interpolation were needed to develop comprehensive estimates of emissions for a few sectors and sink categories, but in most cases, observations and estimates from the year in question were sufficient to generate the necessary data.

This indicator reports trends exactly as they appear in EPA's GHG inventory (U.S. EPA, 2014). The indicator presents emissions data in units of million metric tons of CO_2 equivalents, the conventional unit used in GHG inventories prepared worldwide, because it adjusts for the various global warming potentials (GWPs) of different gases. EPA is required to use the 100-year GWPs documented in the IPCC's Second Assessment Report (SAR) (IPCC, 1996) for the development of the inventory to comply with international reporting standards under the UNFCCC. This requirement ensures that current estimates of aggregate greenhouse gas emissions for 1990 to 2012 are consistent with estimates developed prior to the publication of the IPCC's Third Assessment Report in 2001 and Fourth Assessment Report (AR4) in 2007. Annex 6.1 of the U.S. GHG inventory includes extensive information on GWPs and how they relate to emissions estimates (U.S. EPA, 2014). While greenhouse gas emissions currently presented in this indicator use the SAR GWP values the United States and other developed countries have agreed that, starting in 2015, they will submit annual inventories to the UNFCCC using GWP values from the IPCC's AR4. Thus, the next revision of this indicator will use GWPs from the IPCC AR4.

Figure 1. U.S. Greenhouse Gas Emissions by Gas, 1990–2012

EPA plotted total emissions for each gas, not including the influence of sinks, which would be difficult to interpret in a breakdown by gas. EPA combined the emissions of HFCs, PFCs, and SF_6 into a single category so the magnitude of these emissions would be visible in the graph.

Figure 2. U.S. Greenhouse Gas Emissions and Sinks by Economic Sector, 1990–2012

EPA converted a line graph in the original inventory report (U.S. EPA, 2014) into a stacked area graph showing emissions by economic sector. U.S. territories are treated as a separate sector in the inventory report, and because territories are not an economic sector in the truest sense of the word, they have been excluded from this part of the indicator. Unlike Figure 1, Figure 2 includes sinks below the x-axis.

Figure 3. U.S. Greenhouse Gas Emissions per Capita and per Dollar of GDP, 1990–2012

EPA determined emissions per capita and emissions per unit of real GDP using simple division. In order to show all four trends (population, GDP, emissions per capita, and emissions per unit GDP) on the same scale, EPA normalized each trend to an index value of 100 for the year 1990.

7. Quality Assurance and Quality Control

Quality assurance and quality control (QA/QC) have always been an integral part of the U.S. national system for inventory development. EPA and its partner agencies have implemented a systematic approach to QA/QC for the annual U.S. GHG inventory, following procedures that have been formalized in accordance with a QA/QC plan and the UNFCCC reporting guidelines. Those interested in documentation of the various QA/QC procedures should send such queries to EPA's Climate Change Division (www.epa.gov/climatechange/contactus.html).

Analysis

8. Comparability Over Time and Space

The U.S. GHG emissions data presented in this indicator are comparable over time and space, and the purpose of the inventory is to allow tracking of annual emissions over time. The emissions trend is defined in the inventory as the percentage change in emissions (or removal) estimated for the current year, relative to the emissions (or removal) estimated for the base year (i.e., 1990) inventory estimates. In addition to the estimates of uncertainty associated with the current year's emissions estimates, Annex 7 of the inventory report also presents quantitative estimates of trend uncertainty.

9. Data Limitations

Factors that may impact the confidence, application, or conclusions drawn from this indicator are as follows:

1. This indicator does not yet include emissions of GHGs or other radiatively important substances that are not explicitly covered by the UNFCCC and its subsidiary protocol. Thus, it excludes gases such as those controlled by the Montreal Protocol and its Amendments, including chlorofluorocarbons and hydrochlorofluorocarbons. Although the United States reports the emissions of these substances as part of the U.S. GHG inventory (see Annex 6.2 of U.S. EPA [2013]), the origin of the estimates is fundamentally different from those of the other GHGs, and therefore these emissions cannot be compared directly with the other emissions discussed in this indicator.

2. This indicator does not include aerosols and other emissions that affect radiative forcing and that are not well-mixed in the atmosphere, such as sulfate, ammonia, black carbon, and organic carbon. Emissions of these compounds are highly uncertain and have qualitatively different effects from the six types of emissions in this indicator.

3. This indicator does not include emissions of other compounds—such as carbon monoxide, nitrogen oxides, non-methane volatile organic compounds, and substances that deplete the stratospheric ozone layer—that indirectly affect the Earth's radiative balance (for example, by altering GHG concentrations, changing the reflectivity of clouds, or changing the distribution of heat fluxes).

4. The U.S. GHG inventory does not account for "natural" emissions of GHGs from sources such as wetlands, tundra soils, termites, and volcanoes. These excluded sources are discussed in Annex 5 of the U.S. GHG inventory (U.S. EPA, 2014). The "land use," "land-use change," and "forestry" categories in U.S. EPA (2014) do include emissions from changes in the forest inventory due to fires, harvesting, and other activities, as well as emissions from agricultural soils.

10. Sources of Uncertainty

Some estimates, such as those for CO_2 emissions from energy-related activities and cement processing, are considered to have low uncertainties. For some other categories of emissions, however, lack of data or an incomplete understanding of how emissions are generated increases the uncertainty of the estimates presented.

Recognizing the benefit of conducting an uncertainty analysis, the UNFCCC reporting guidelines follow the recommendations of IPCC (2000) and require that countries provide single point uncertainty estimates for many sources and sink categories. The U.S. GHG inventory (U.S. EPA, 2014) provides a qualitative discussion of uncertainty for all sources and sink categories, including specific factors affecting the uncertainty surrounding the estimates. Most sources also have a quantitative uncertainty assessment in accordance with the new UNFCCC reporting guidelines. Thorough discussion of these points can be found in U.S. EPA (2014). Annex 7 of the inventory publication is devoted entirely to uncertainty in the inventory estimates.

For a general idea of the degree of uncertainty in U.S. emissions estimates, WRI (2013) provides the following information: "Using IPCC Tier 2 uncertainty estimation methods, EIA (2002) estimated uncertainties surrounding a simulated mean of CO_2 (-1.4% to 1.3%), CH_4 (-15.6% to 16%), and N_2O (-53.5% to 54.2%). Uncertainty bands appear smaller when expressed as percentages of total estimated emissions: CO_2 (-0.6% to 1.7%), CH_4 (-0.3% to 3.4%), and N_2O (-1.9% to 6.3%)."

EPA is investigating studies and sources of uncertainty in estimates of methane emissions from oil and gas development. For example, EPA is currently seeking stakeholder feedback on how information from such measurement studies can be used to update inventory estimates. Some factors for consideration include whether measurements taken are representative of all natural gas producing areas in the United States, what activities were taking place at the time of measurement (general operating conditions or high-emission venting events), and how such measurements can inform emission factors and activity data used to calculate national emissions.

Overall, these sources of uncertainty are not expected to have a considerable impact on this indicator's conclusions. Even considering the uncertainties of omitted sources and lack of precision in known and estimated sources, this indicator provides a generally accurate picture of aggregate trends in GHG emissions over time, and hence the overall conclusions inferred from the data are solid. The U.S. GHG inventory represents the most comprehensive and reliable data set available to characterize GHG emissions in the United States.

11. Sources of Variability

Within each sector (e.g., electricity generation), GHG emissions can vary considerably across the individual point sources, and many factors contribute to this variability (e.g., different production levels, fuel type, air pollution controls). EPA's inventory methods account for this variability among individual emissions sources.

12. Statistical/Trend Analysis

This indicator presents a time series of national emissions estimates. No special statistical techniques or analyses were used to characterize the long-term trends or their statistical significance.

References

IPCC (Intergovernmental Panel on Climate Change). 1996. Climate change 1995: The science of climate change. Cambridge, United Kingdom: Cambridge University Press.

IPCC (Intergovernmental Panel on Climate Change). 2000. Good practice guidance and uncertainty management in national greenhouse gas inventories. www.ipcc-nggip.iges.or.jp/public/gp/english.

IPCC (Intergovernmental Panel on Climate Change). 2006. IPCC guidelines for national greenhouse gas inventories. www.ipcc-nggip.iges.or.jp/public/2006gl/index.html.

IPCC (Intergovernmental Panel on Climate Change). 2013. Climate change 2013: The physical science basis. Working Group I contribution to the IPCC Fifth Assessment Report. Cambridge, United Kingdom: Cambridge University Press. www.ipcc.ch/report/ar5/wg1.

U.S. EPA. 2014. Inventory of U.S. greenhouse gas emissions and sinks: 1990–2012. USEPA #EPA 430-R-14-003. www.epa.gov/climatechange/ghgemissions/usinventoryreport.html.

WRI (World Resources Institute). 2013. CAIT 2.0: Country greenhouse gas sources and methods. http://cait2.wri.org/docs/CAIT2.0_CountryGHG_Methods.pdf.

Global Greenhouse Gas Emissions

Identification

1. Indicator Description

This indicator describes emissions of greenhouse gases (GHGs) worldwide since 1990. This indicator is highly relevant to climate change because greenhouse gases from human activities are the primary driver of observed climate change since the mid-20th century (IPCC, 2013). Tracking GHG emissions worldwide provides a context for understanding the United States' role in addressing climate change.

Components of this indicator include:

- Global GHG emissions by gas (Figure 1)
- Global GHG emissions by sector (Figure 2)
- Global GHG emissions by regions of the world (Figure 3)

2. Revision History

April 2010:   Indicator posted.
December 2011:   Updated with new and revised data points.
April 2012:   Updated with revised data points.
May 2014:   Updated with revised data points; added land-use change and forestry data.

Data Sources

3. Data Sources

This indicator is based on data from the World Resources Institute's (WRI's) Climate Analysis Indicators Tool (CAIT), a database of anthropogenic sources and sinks of GHGs worldwide. CAIT has compiled data from a variety of GHG emissions inventories. In general, a GHG emissions inventory consists of estimates derived from direct measurements, aggregated national statistics, and validated models. EPA obtained data from CAIT Version 2.0.

CAIT compiles data from a variety of other databases and inventories, including:

- International Energy Agency (IEA) data on carbon dioxide (CO_2) emissions from combustion.
- EPA's estimates of global emissions of non-CO_2 gases.
- Estimates of CO_2 emissions from land-use change and forestry (LUCF), as compiled by the Food and Agriculture Organization of the United Nations (FAO).
- Additional data from the U.S. Carbon Dioxide Information Analysis Center (CDIAC) and U.S. Energy Information Administration (EIA) to fill gaps.

Other global emissions estimates—such as the estimates published by the Intergovernmental Panel on Climate Change (e.g., IPCC, 2013)—are based on many of the same sources.

Note that as a condition to EPA's presentation of FAO data in non-United Nations contexts, FAO asserts that it does not endorse any views, products, or services associated with the presentation of its data.

EPA uses CAIT as the primary data source of this indicator for several reasons, including:

- WRI compiles datasets exclusively from peer-reviewed and authoritative sources, which are easily accessible through CAIT.

- CAIT allows for consistent and routine updates of this indicator, whereas data compiled from other sources (e.g., periodic assessment reports from the Intergovernmental Panel on Climate Change [IPCC]) may be superseded as soon as CAIT's underlying data sources have published newer numbers.

- CAIT relies exclusively on EPA's global estimates for non-CO_2 gases.

- Global estimates from CAIT (excluding LUCF) are comparable with other sources of global GHG data (e.g., the European Commission's Emission Database for Global Atmospheric Research [EDGAR]).

4. Data Availability

All indicator data can be obtained from the WRI CAIT database at: http://cait.wri.org. CAIT includes documentation that describes the various data fields and their sources.

Many of the original data sources are publicly available. For information on all the sources used to populate the CAIT database by country, by gas, and by source or sink category, see WRI (2014). Data for this particular indicator were compiled by WRI largely from the following sources:

- Boden et al. (2013)
- EIA (2013)
- FAO (2014)
- IEA (2013)
- U.S. EPA (2012)

See: http://cait.wri.org for a list of which countries belong to each of the regions shown in Figure 3.

Methodology

5. Data Collection

This indicator focuses on emissions of the six compounds or groups of compounds currently covered by agreements under the United Nations Framework Convention on Climate Change (UNFCCC). These compounds are CO_2, methane (CH_4), nitrous oxide (N_2O), selected hydrofluorocarbons (HFCs), selected perfluorocarbons (PFCs), and sulfur hexafluoride (SF_6). This indicator presents emissions data in units of million metric tons of CO_2 equivalents, the conventional unit used in GHG inventories prepared worldwide, because it adjusts for the various global warming potentials (GWPs) of different gases.

The data originally come from a variety of GHG inventories. Some have been prepared by national governments; others by international agencies. Data collection techniques (e.g., survey design) vary depending on the source or parameter. For example, FAO is acknowledged as an authoritative source of land-use-related emissions data because they are able to estimate deforestation patterns with help from satellite imagery (Houghton et al., 2012). Although the CAIT database is intended to be comprehensive, the organizations that develop inventories are continually working to improve their understanding of emissions sources and how best to quantify them.

Inventories often use some degree of extrapolation and interpolation to develop comprehensive estimates of emissions in a few sectors and sink categories, but in most cases, observations and estimates from the year in question were sufficient to generate the necessary data.

GHG inventories are not based on any one specific sampling plan, but documents are available that describe how most inventories have been constructed. For example, U.S. EPA (2014) describes all the procedures used to estimate GHG emissions for EPA's annual U.S. inventory. See the IPCC's GHG inventory guidelines (IPCC, 2006) and IPCC's *Good Practice Guidance and Uncertainty Management in National Greenhouse Gas Inventories* (IPCC, 2000) for additional guidance that many countries and organizations follow when constructing GHG inventories.

6. Indicator Derivation

This indicator reports selected metrics from WRI's CAIT database, which compiles data from the most reputable GHG inventories around the world. WRI's website (http://cait2.wri.org/faq.html) provides an overview of how the CAIT database was constructed, and WRI (2014) describes the data sources and methods used to populate the database. WRI's main role is to assemble data from other sources, all of which have been critically reviewed. As a result, the totals reported in CAIT are consistent with other compilations, such as a European tool called EDGAR (http://edgar.jrc.ec.europa.eu/index.php), which has been cited in reports by IPCC. EDGAR and CAIT use many of the same underlying data sources.

The most comprehensive estimates are available beginning in 1990. Global emissions estimates for CO_2 are available annually through 2011, while global estimates for gases other than CO_2 are available only at five-year intervals through 2010. Thus, Figures 1 and 2 (which show all GHGs) plot values for 1990, 1995, 2000, 2005, and 2010. WRI and EPA did not attempt to interpolate estimates for the interim years.

All three figures in this indicator include emissions due to international transport (i.e., aviation and maritime bunker fuel). These emissions are not included in the U.S. Greenhouse Gas Emissions indicator because they are international by nature, and not necessarily reflected in individual countries' emissions inventories.

Figures 1 and 2 include estimates of emissions associated with LUCF. Figure 3 excludes LUCF because it focuses on gross emissions by region.

The indicator presents emissions data in units of million metric tons of CO_2 equivalents, which are conventionally used in GHG inventories prepared worldwide because they adjust for the various global warming potentials (GWPs) of different gases. This analysis uses the 100-year GWPs that are documented in the IPCC's Second Assessment Report (SAR) (IPCC, 1996). This choice arises because CAIT's data for non-CO_2 gases come from an EPA global compilation, and EPA uses SAR GWPs in products such as the annual U.S. GHG inventory because the Agency is required to do so to comply with international reporting standards under the UNFCCC. This requirement ensures that current estimates of aggregate greenhouse gas emissions are consistent with estimates developed prior to the publication of the IPCC's Third Assessment Report in 2001 and Fourth Assessment Report (AR4) in 2007. While greenhouse gas emissions currently presented in this indicator use the SAR GWP values, this will be the final time the SAR GWP values will be used in the annual U.S. GHG inventory and related reports (e.g., EPA's global non-CO_2 data compilation). The United States and other developed countries have agreed to submit annual inventories in 2015 and future years to the UNFCCC using GWP values from the IPCC's AR4, which will replace the current use of SAR GWP values. Thus, future editions of this indicator will use GWPs from the IPCC AR4.

Figure 1. Global Greenhouse Gas Emissions by Gas, 1990–2010

EPA plotted total emissions for each gas, combining the emissions of HFCs, PFCs, and SF_6 into a single category so the magnitude of these emissions would be visible in the graph. EPA formatted the graph as a series of stacked columns instead of a continuous stacked area because complete estimates for all gases are available only every five years, and it would be misleading to suggest that information is known about trends in the interim years.

Figure 2. Global Greenhouse Gas Emissions by Sector, 1990–2010

EPA plotted total GHG emissions by IPCC sector. IPCC sectors are different from the sectors used in Figure 2 of the U.S. Greenhouse Gas Emissions indicator, which uses an economic sector breakdown that is not available on a global scale. EPA formatted the graph as a series of stacked columns instead of a continuous stacked area because complete estimates for all gases are available only every five years, and it would be misleading to suggest that information is known about trends in the interim years.

Figure 3. Global Carbon Dioxide Emissions by Region, 1990–2011

In order to show data at more than four points in time, EPA elected to display emissions by region for CO_2 only, as CO_2 emissions estimates are available with annual resolution. EPA performed basic calculations to ensure that no emissions were double-counted across the regions. Specifically, EPA subtracted U.S. totals from North American totals, leaving "Other North America" (which also includes Central America and the Caribbean) as a separate category. For "Africa and the Middle East," EPA combined North Africa and Sub-Saharan Africa.

Indicator Development

In the course of developing and revising this indicator, EPA considered data from a variety of sources, including WRI's CAIT (http://cait.wri.org) and EDGAR (http://edgar.jrc.ec.europa.eu/index.php). EPA compared data obtained from CAIT and EDGAR for global carbon dioxide emissions, and found the two data sources were highly comparable for global estimates of non-LUCF data, with differences of less than 2 percent for all years.

For the purposes of CAIT, WRI essentially serves as a secondary compiler of global emissions data, drawing on internationally recognized inventories from government agencies and using extensively peer-reviewed data sets. EPA has determined that WRI does not perform additional interpolations on the data, but rather makes certain basic decisions in order to allocate emissions to certain countries (e.g., in the case of historical emissions from Soviet republics). These methods are described in CAIT's supporting documentation, which EPA carefully reviewed to assure the credibility of the source.

Previous versions of EPA's indicator excluded LUCF because of its relatively large uncertainty and because of a time lag at the original source that caused LUCF estimates to be up to five years behind the other source categories in this indicator. However, the scientific community has developed a better understanding of LUCF uncertainties (see Section 10) and FAO now provides timely global LUCF estimates based on improved methods. As Houghton et al. (2012) note, "Better reporting of deforestation rates by the FAO [due to the inclusion of satellite data] has narrowed the range of estimates cited by Houghton (2005) and the IPCC (2007) and is likely to reduce the uncertainty still more in the future." Thus, EPA added LUCF to this indicator in 2014.

Additionally, FAO activity data are the default data to which the UNFCCC expert review teams compare country-specific data when performing GHG inventory reviews under the Convention. If the data a country uses to perform its calculations differ significantly from FAO data, the country needs to explain the discrepancy.

7. Quality Assurance and Quality Control

Quality assurance and quality control (QA/QC) documentation is not explicitly provided with the full CAIT database, but many of the contributing sources have documented their QA/QC procedures. For example, EPA and its partner agencies have implemented a systematic approach to QA/QC for the annual U.S. GHG inventory, following procedures that have recently been formalized in accordance with a QA/QC plan and the UNFCCC reporting guidelines. Those interested in documentation of the various QA/QC procedures for the U.S. inventory should send such queries to EPA's Climate Change Division (www.epa.gov/climatechange/contactus.html). QA/QC procedures for other sources can generally be found in the documentation that accompanies the sources cited in Section 4.

Analysis

8. Comparability Over Time and Space

Some inventories have been prepared by national governments; others by international agencies. Data collection techniques (e.g., survey design) vary depending on the source or parameter. To the extent possible, inventories follow a consistent set of best practice guidelines described in IPCC (2000, 2006).

9. Data Limitations

Factors that may impact the confidence, application, or conclusions drawn from this indicator are as follows:

1. This indicator does not yet include emissions of GHGs or other radiatively important substances that are not explicitly covered by the UNFCCC and its subsidiary protocol. Thus, it excludes gases such as those controlled by the Montreal Protocol and its Amendments, including chlorofluorocarbons and hydrochlorofluorocarbons. Although some countries report emissions of these substances, the origin of the estimates is fundamentally different from those of other

GHGs, and therefore these emissions cannot be compared directly with the other emissions discussed in this indicator.

2. This indicator does not include aerosols and other emissions that affect radiative forcing and that are not well-mixed in the atmosphere, such as sulfate, ammonia, black carbon, and organic carbon. Emissions of these compounds are highly uncertain and have qualitatively different effects from the six types of emissions in this indicator.

3. This indicator does not include emissions of other compounds—such as carbon monoxide, nitrogen oxides, nonmethane volatile organic compounds, and substances that deplete the stratospheric ozone layer—which indirectly affect the Earth's radiative balance (for example, by altering GHG concentrations, changing the reflectivity of clouds, or changing the distribution of heat fluxes).

4. The LUCF component of this indicator is limited to the CO_2 estimates available from FAO, which cover emissions and sinks associated with forest land, grassland, cropland, and biomass burning. FAO excludes wetlands, settlements, and "other" categories, but these sources/sinks are relatively small on a global scale, and FAO's four categories constitute a large majority of LUCF emissions and sinks. This indicator also does not include non-CO_2 LUCF emissions and sinks, which are estimated to be much smaller than CO_2 totals.

5. This indicator does not account for "natural" emissions of GHGs, such as from wetlands, tundra soils, termites, and volcanoes.

6. Global emissions data for non-CO_2 GHGs are only available at five-year intervals. Thus, Figures 1 and 2 show data for just five points in time: 1990, 1995, 2000, 2005, and 2010.

10. Sources of Uncertainty

In general, all emissions estimates will have some inherent uncertainty. Estimates of CO_2 emissions from energy-related activities and cement processing are often considered to have the lowest uncertainties, but even these data can have errors as a result of uncertainties in the numbers from which they are derived, such as national energy use data. In contrast, estimates of emissions associated with land-use change and forestry may have particularly large uncertainties. As Ito et al. (2008) explain, "Because there are different sources of errors at the country level, there is no easy reconciliation of different estimates of carbon fluxes at the global level. Clearly, further work is required to develop data sets for historical land cover change areas and models of biogeochemical changes for an accurate representation of carbon uptake or emissions due to [land-use change]." Houghton et al. (2012) reviewed 13 different estimates of global emissions from land use, land-use change, and forestry. They estimated an overall error of ±500 million metric tons of carbon per year. This estimate represents an improvement in understanding LUCF, but still results in a larger uncertainty than can be found in other sectors.

The Modeling and Assessment of Contributions of Climate Change (MATCH) group has thoroughly reviewed a variety of global emissions estimates to characterize their uncertainty. A summary report and detailed articles are available on the MATCH website at: www.match-info.net.

For specific information about uncertainty, users should refer to documentation from the individual data sources cited in Section 4. Uncertainty estimates are available from the underlying national inventories in some cases, in part because the UNFCCC reporting guidelines follow the recommendations of IPCC (2000) and require countries to provide single point uncertainty estimates for many sources and sink categories. For example, the U.S. GHG inventory (U.S. EPA, 2014) provides a qualitative discussion of uncertainty for all sources and sink categories, including specific factors affecting the uncertainty of the estimates. Most sources also have a quantitative uncertainty assessment, in accordance with the new UNFCCC reporting guidelines. Thorough discussion of these points can be found in U.S. EPA (2014). Annex 7 of EPA's inventory publication is devoted entirely to uncertainty in the inventory estimates. Uncertainties are expected to be greater in estimates from developing countries, due in some cases to varying quality of underlying activity data and uncertain emissions factors. Uncertainties are generally greater for non-CO_2 gases than for CO_2.

Uncertainty is not expected to have a considerable impact on this indicator's conclusions. Uncertainty is indeed present in all emissions estimates, in some cases to a great degree—especially for LUCF and for non-CO_2 gases in developing countries. At an aggregate global scale, however, this indicator accurately depicts the overall direction and magnitude of GHG emissions trends over time, and hence the overall conclusions inferred from the data are reasonable.

The FAO data set has certain limitations that result from the application of Tier 1 methods and from uncertainties in the underlying data. Specific estimates of the uncertainty are not readily available from FAO, and it would be complicated and speculative to provide uncertainty bounds around these data.

11. Sources of Variability

On a national or global scale, year-to-year variability in GHG emissions can arise from a variety of factors, such as economic conditions, fuel prices, and government actions. Overall, variability is not expected to have a considerable impact on this indicator's conclusions.

12. Statistical/Trend Analysis

This indicator does not report on the slope of the apparent trends in global GHG emissions, nor does it calculate the statistical significance of these trends. The "Key Points" describe percentage change between 1990 and the most recent year of data—an endpoint-to-endpoint comparison, not a trend line of best fit.

References

Boden, T.A., G. Marland, and R. J. Andres. 2013. Global, regional, and national fossil fuel CO_2 emissions. Carbon Dioxide Information Analysis Center, Oak Ridge National Laboratory, U.S. Department of Energy. http://cdiac.ornl.gov/trends/emis/overview_2010.html.

EIA (U.S. Energy Information Administration). 2013. International energy statistics. http://www.eia.gov/countries/data.cfm.

FAO (Food and Agriculture Organization). 2014. FAOSTAT: Emissions—land use. Accessed May 2014. http://faostat3.fao.org/faostat-gateway/go/to/download/G2/*/E.

Houghton, R.A., G.R. van der Werf, R.S. DeFries, M.C. Hansen, J.I. House, C. Le Quéré, J. Pongratz, and N. Ramankutty. 2012. Chapter G2: Carbon emissions from land use and land-cover change. Biogeosciences Discuss. 9(1):835–878.

IEA (International Energy Agency). 2013. CO_2 emissions from fuel combustion (2013 edition). Paris, France: OECD/IEA. http://data.iea.org/ieastore/statslisting.asp.

IPCC (Intergovernmental Panel on Climate Change). 1996. Climate change 1995: The science of climate change. Cambridge, United Kingdom: Cambridge University Press.

IPCC (Intergovernmental Panel on Climate Change). 2000. Good practice guidance and uncertainty management in national greenhouse gas inventories. www.ipcc-nggip.iges.or.jp/public/gp/english.

IPCC (Intergovernmental Panel on Climate Change). 2006. IPCC guidelines for national greenhouse gas inventories. www.ipcc-nggip.iges.or.jp/public/2006gl/index.html.

IPCC (Intergovernmental Panel on Climate Change). 2013. Climate change 2013: The physical science basis. Working Group I contribution to the IPCC Fifth Assessment Report. Cambridge, United Kingdom: Cambridge University Press. www.ipcc.ch/report/ar5/wg1.

Ito, A., J.E. Penner, M.J. Prather, C.P. de Campos, R.A. Houghton, T. Kato, A.K. Jain, X. Yang, G.C. Hurtt, S. Frolking, M.G. Fearon, L.P. Chini, A. Wang, and D.T. Price. 2008. Can we reconcile differences in estimates of carbon fluxes from land-use change and forestry for the 1990s? Atmos. Chem. Phys. 8:3291–3310.

U.S. EPA. 2012. Global anthropogenic non-CO_2 greenhouse gas emissions: 1990–2030. www.epa.gov/climatechange/EPAactivities/economics/nonco2projections.html.

U.S. EPA. 2014. Inventory of U.S. greenhouse gas emissions and sinks: 1990–2012. www.epa.gov/climatechange/ghgemissions/usinventoryreport.html.

WRI (World Resources Institute). 2014. CAIT 2.0: Country greenhouse gas sources and methods. http://cait2.wri.org/docs/CAIT2.0_CountryGHG_Methods.pdf.

Atmospheric Concentrations of Greenhouse Gases

Identification

1. Indicator Description

This indicator describes how the levels of major greenhouse gases (GHGs) in the atmosphere have changed over geological time and in recent years. Changes in atmospheric GHGs, in part caused by human activities, affect the amount of energy held in the Earth-atmosphere system and thus affect the Earth's climate. This indicator is highly relevant to climate change because greenhouse gases from human activities are the primary driver of observed climate change since the mid-20th century (IPCC, 2013).

Components of this indicator include:

- Global atmospheric concentrations of carbon dioxide over time (Figure 1).
- Global atmospheric concentrations of methane over time (Figure 2).
- Global atmospheric concentrations of nitrous oxide over time (Figure 3).
- Global atmospheric concentrations of selected halogenated gases over time (Figure 4).
- Global atmospheric concentrations of ozone over time (Figure 5).

2. Revision History

April 2010:   Indicator posted.
December 2011:   Updated with data through 2010.
May 2012:   Updated with data through 2011.
July 2012:   Added nitrogen trifluoride to Figure 4.
August 2013:   Updated indicator on EPA's website with data through 2012.
December 2013:   Added Figure 5 to show trends in ozone.
May 2014:   Updated with data through 2012 (full year) for Figure 4 and through 2013 for other figures.

Data Sources

3. Data Sources

Ambient concentration data used to develop this indicator were taken from the following sources:

Figure 1. Global Atmospheric Concentrations of Carbon Dioxide Over Time

- EPICA Dome C and Vostok Station, Antarctica: approximately 796,562 BC to 1813 AD—Lüthi et al. (2008).
- Law Dome, Antarctica, 75-year smoothed: approximately 1010 AD to 1975 AD—Etheridge et al. (1998).
- Siple Station, Antarctica: approximately 1744 AD to 1953 AD—Neftel et al. (1994).

- Mauna Loa, Hawaii: 1959 AD to 2013 AD—NOAA (2014a).
- Barrow, Alaska: 1974 AD to 2012 AD; Cape Matatula, American Samoa: 1976 AD to 2012 AD; South Pole, Antarctica: 1976 AD to 2012 AD—NOAA (2014c).
- Cape Grim, Australia: 1992 AD to 2006 AD; Shetland Islands, Scotland: 1993 AD to 2002 AD—Steele et al. (2007).
- Lampedusa Island, Italy: 1993 AD to 2000 AD—Chamard et al. (2001).

Figure 2. Global Atmospheric Concentrations of Methane Over Time

- EPICA Dome C, Antarctica: approximately 797,446 BC to 1937 AD—Loulergue et al. (2008).
- Law Dome, Antarctica: approximately 1008 AD to 1980 AD—Etheridge et al. (2002).
- Cape Grim, Australia: 1984 AD to 2013 AD—NOAA (2014d).
- Mauna Loa, Hawaii: 1987 AD to 2013 AD—NOAA (2014e).
- Shetland Islands, Scotland: 1993 AD to 2001 AD—Steele et al. (2002).

Figure 3. Global Atmospheric Concentrations of Nitrous Oxide Over Time

- EPICA Dome C, Antarctica: approximately 796,475 BC to 1937 AD—Schilt et al. (2010).
- Antarctica: approximately 1903 AD to 1976 AD—Battle et al. (1996).
- Cape Grim, Australia: 1979 AD to 2012 AD—AGAGE (2014b).
- South Pole, Antarctica: 1998 AD to 2013 AD; Barrow, Alaska: 1999 AD to 2013 AD; Mauna Loa, Hawaii: 2000 AD to 2013 AD—NOAA (2014f).

Figure 4. Global Atmospheric Concentrations of Selected Halogenated Gases, 1978–2012

Global average atmospheric concentration data for selected halogenated gases were obtained from the following sources:

- National Oceanic and Atmospheric Administration (NOAA, 2013) for halon-1211.
- Arnold (2013) for nitrogen trifluoride.
- Advanced Global Atmospheric Gases Experiment (AGAGE, 2014a) for all other species shown.

A similar figure based on Advanced Global Atmospheric Gases Experiment (AGAGE) and National Oceanic and Atmospheric Administration (NOAA) data appears in the Intergovernmental Panel on Climate Change's (IPCC's) Fifth Assessment Report (see Figure 2.4 in IPCC, 2013).

Figure 5. Global Atmospheric Concentrations of Ozone , 1979–2013

Ozone data were obtained from several National Aeronautics and Space Administration (NASA) sources:

- The Solar Backscatter Ultraviolet (SBUV) merged ozone data set (NASA, 2014a) for total ozone.
- The Tropospheric Ozone Residual (TOR) (NASA, 2013) and Ozone Monitoring Instrument (OMI) Level 2 (NASA, 2014b) data sets for tropospheric ozone.

4. Data Availability

The data used to develop Figures 1, 2, 3, and 5 of this indicator are publicly available and can be accessed from the references listed in Section 3. There are no known confidentiality issues.

Data for all of the halogenated gases in Figure 4, with the exception of halon-1211 and nitrogen trifluoride, were downloaded from the AGAGE website at:
http://agage.eas.gatech.edu/data_archive/global_mean. Additional historical monthly data for some of these gases were provided in spreadsheet form by Dr. Ray Wang of the AGAGE project team (AGAGE, 2011). Bimonthly data for halon-1211 were provided in spreadsheet form by Dr. Stephen Montzka of NOAA (NOAA, 2014b). NOAA's website (www.esrl.noaa.gov/gmd/hats) provides access to underlying station-specific data and selected averages, but does not provide the global averages that are shown in Figure 4. Nitrogen trifluoride data are based on measurements that were originally published in Arnold et al. (2013) and subsequently updated by the lead author (Arnold, 2013).

In the event of a time lag in data being posted to the websites cited in Section 3, updated data can be obtained by contacting the original data provider (e.g., NOAA or NASA staff).

Methodology

5. Data Collection

This indicator shows trends in atmospheric concentrations of several major GHGs that enter the atmosphere at least in part because of human activities: carbon dioxide (CO_2), methane (CH_4), nitrous oxide (N_2O), selected halogenated gases, and ozone.

Figures 1, 2, 3, and 4. Global Atmospheric Concentrations of Carbon Dioxide, Methane, Nitrous Oxide, and Selected Halogenated Gases Over Time

Figures 1, 2, 3, and 4 aggregate comparable, high-quality data from individual studies that each focused on different locations and time frames. Data since the mid-20th century come from global networks that use standard monitoring techniques to measure the concentrations of gases in the atmosphere. Older measurements of atmospheric concentrations come from ice cores—specifically, measurements of gas concentrations in air bubbles that were trapped in ice at the time the ice was formed. Scientists have spent years developing and refining methods of measuring gases in ice cores as well as methods of dating the corresponding layers of ice to determine their age. Ice core measurements are a widely used method of reconstructing the composition of the atmosphere before the advent of direct monitoring techniques.

This indicator presents a compilation of data generated by numerous sampling programs. The citations listed in Section 3 describe the specific approaches taken by each program. Gases are measured by mole fraction relative to dry air.

CO_2, CH_4, N_2O, and most of the halogenated gases presented in this indicator are considered to be well-mixed globally, due in large part to their long residence times in the atmosphere. Thus, while measurements over geological time tend to be available only for regions where ice cores can be collected (e.g., the Arctic and Antarctic regions), these measurements are believed to adequately represent concentrations worldwide. Recent monitoring data have been collected from a greater variety of locations, and the results show that concentrations and trends are indeed very similar throughout the world, although relatively small variations can be apparent across different locations.

Most of the gases shown in Figure 4 have been measured around the world numerous times per year. One exception is nitrogen trifluoride, which is an emerging gas of concern for which measurements are not yet widespread. The curve for nitrogen trifluoride in Figure 4 is based on samples collected in Australia and California between 1995 and 2010, plus samples collected approximately monthly in California since 2010. Measurements of air samples collected before 1995 have also been made, but they are not shown in this figure because larger gaps in time exist between these measurements.

Nitrogen trifluoride was measured by the Medusa gas chromatography with mass spectrometry (GCMS) system, with refinements described in Weiss et al. (2008), Arnold et al. (2012), and Arnold et al. (2013). Mole fractions of the other halogenated gases were collected by AGAGE's Medusa GCMS system, or similar methods employed by NOAA.

Figure 5. Global Atmospheric Concentrations of Ozone, 1979–2013

Unlike the gases in Figures 1, 2, 3, and 4, which are measured as atmospheric concentrations near ground level, Figure 5 describes the total "thickness" of ozone in the Earth's atmosphere. This measurement is called total column ozone, and it is typically measured in Dobson units. One Dobson unit represents a layer of gas that would be 10 micrometers (μm) thick under standard temperature and pressure (0°C/32°F and 0.987 atmospheres of air pressure).

Atmospheric ozone concentrations for this indicator are based on measurements by three sets of satellite instruments:

- **SBUV.** The SBUV observing system consists of a series of instruments aboard nine satellites that have collectively covered the period from 1970 to present, except for a gap from 1972 to 1978. The SBUV measures the total ozone profile from the Earth's surface to the upper edge of the atmosphere (total column ozone) by analyzing solar backscatter radiation, which is the visible light and ultraviolet radiation that the Earth's atmosphere reflects back to space. This instrument can be used to determine the amount of ozone in each of 21 discrete layers of the atmosphere, which are then added together to get total column ozone. For a table of specific SBUV satellite instruments and the time periods they cover, see: http://acdb-ext.gsfc.nasa.gov/Data_services/merged/instruments.html. A new instrument, the Ozone Mapping Profiler Suite (OMPS) Nadir Profiler, will continue the SBUV series. Although instrument design has improved over time, the basic principles of the measurement technique and processing algorithm remain the same, lending consistency to the record. For more information about the SBUV data set and how it was collected, see McPeters et al. (2013) and the references listed at: http://acdb-ext.gsfc.nasa.gov/Data_services/merged/index.html.

- **Total Ozone Mapping Spectrometer (TOMS).** TOMS instruments have flown on four satellite missions that collectively cover the period from 1978 to 2005, with the exception of a period from late 1994 to early 1996 when no TOMS instrument was in orbit. Like the SBUV, the TOMS measured total ozone in the Earth's atmosphere by analyzing solar backscatter radiation. For more information about TOMS missions and instrumentation, see: http://disc.sci.gsfc.nasa.gov/acdisc/TOMS.

- **Aura OMI and Microwave Limb Sounder (MLS)**. The Aura satellite was launched in 2004, and its instruments (including OMI and MLS) were still collecting data as of 2014. The OMI instrument measures total column ozone by analyzing solar backscatter radiation. In contrast, the MLS measures emissions of microwave radiation from the Earth's atmosphere. This method allows the MLS to characterize the temperature and composition of specific layers of the atmosphere, including the amount of ozone within the stratosphere. To learn more about the Aura mission and its instruments, visit: http://aura.gsfc.nasa.gov and: http://aura.gsfc.nasa.gov/scinst/index.html.

The instruments described above have flown on polar-orbiting satellites, which collect measurements that cover the entire surface of the Earth. However, for reasons of accuracy described in Section 9, this indicator is limited to data collected between 50°N and 50°S latitude. Solar backscatter measurements are restricted to daytime, when the sun is shining on a particular part of the Earth and not too low in the sky (i.e., avoiding measurements near sunrise or sunset).

6. Indicator Derivation

EPA obtained and compiled data from various GHG measurement programs and plotted these data in graphs. No attempt was made to project concentrations backward before the beginning of the ice core record (or the start of monitoring, in the case of Figures 4 and 5) or forward into the future.

Figures 1, 2, and 3. Global Atmospheric Concentrations of Carbon Dioxide, Methane, and Nitrous Oxide Over Time

Figures 1, 2, and 3 plot data at annual or multi-year intervals; with ice cores, consecutive data points are often spaced many years apart. EPA used the data exactly as reported by the organizations that collected them, with the following exceptions:

- Some of the recent time series for CO_2, CH_4, and N_2O consisted of monthly measurements. EPA averaged these monthly measurements to arrive at annual values to plot in the graphs. A few years did not have data for all 12 months. If at least nine months of data were present in a given year, EPA averaged the available data to arrive at an annual value. If fewer than nine monthly measurements were available, that year was excluded from the graph.

- Some ice core records were reported in terms of the age of the sample or the number of years before present. EPA converted these dates into calendar years.

- A few ice core records had multiple values at the same point in time (i.e., two or more different measurements for the same year). These values were generally comparable and never varied by more than 4.8 percent. In such cases, EPA averaged the values to arrive at a single atmospheric concentration per year.

Figures 1, 2, and 3 present a separate line for each data series or location where measurements were collected. No methods were used to portray data for locations other than where measurements were made. However, the indicator does imply that the values in the graphs represent global atmospheric concentrations—an appropriate assumption because the gases covered by this indicator have long residence times in the atmosphere and are considered to be well-mixed. In the indicator text, the Key

Points refer to the concentration for the most recent year available. If data were available for more than one location, the text refers to the average concentration across these locations.

Figure 4. Global Atmospheric Concentrations of Selected Halogenated Gases, 1978–2012

Figure 4 plots data at sub-annual intervals (i.e., several data points per year). EPA used the data exactly as reported by the organizations that collected them, with one exception, for nitrogen trifluoride. Although measurements have been made of nitrogen trifluoride in air samples collected before 1995, EPA elected to start the nitrogen trifluoride time series at 1995 because of large time gaps between measurements prior to 1995. The individual data point for 1995 is not shown because it falls below the scale of Figure 4's y-axis.

Figure 4 presents one trend line for each halogenated gas, and these lines represent average concentrations across all measurement sites worldwide. These data represent monthly average mole fractions for each species, with two exceptions: halon-1211 data are only available at two-month intervals, and nitrogen trifluoride measurements were converted into global annual average mole fractions using a model described in Arnold et al. (2013). This update of nitrogen trifluoride represents a change from the version of this indicator that EPA initially published in December 2012. At the time of the December 2012 version, modeled global annual average mole fractions had not yet been published in the literature, so EPA's indicator instead relied upon individual measurements of nitrogen trifluoride that were limited to the Northern Hemisphere.

Data are available for additional halogenated species, but to make the most efficient use of the space available, EPA selected a subset of gases that are relatively common, have several years of data available, show marked growth trends (either positive or negative), and/or collectively represent most of the major categories of halogenated gases. The inclusion of nitrogen trifluoride here is based on several factors. Like perfluoromethane (PFC-14 or CF_4), perfluoroethane (PFC-116 or C_2F_6), and sulfur hexafluoride, nitrogen trifluoride is a widely produced, fully fluorinated gas with a very high 100-year global warming potential (17,200) and a long atmospheric lifetime (740 years). Nitrogen trifluoride has experienced a rapid increase in emissions (i.e., more than 10 percent per year) due to its use in manufacturing semiconductors, flat screen displays, and thin film solar cells. It began to replace perfluoroethane in the electronics industry in the late 1990s.

To examine the possible influence of phase-out and substitution activities under the Montreal Protocol on Substances That Deplete the Ozone Layer, EPA divided Figure 4 into two panels: one for substances officially designated as "ozone-depleting" and one for all other halogenated gases.

Figure 5. Global Atmospheric Concentrations of Ozone, 1979–2013

NASA converted the satellite measurements into meaningful data products using the following methods:

- Data from all SBUV instruments were processed using the Version 8.6 algorithm (Bhartia et al., 2012; Kramarova et al., 2013b). The resulting data set indicates the amount of ozone in each of 21 distinct atmospheric layers, in Dobson units.

- NASA developed the TOR data set, which represents ozone in the troposphere only. They did so by starting with total column ozone measurements from TOMS and SBUV, then subtracting the portion that could be attributed to the stratosphere. NASA developed this method using information about the height of the tropopause (the boundary between the troposphere and the stratosphere) over time and space, stratosphere-only ozone measurements from the Stratospheric Aerosol and Gas Experiment (SAGE) instrument that flew on some of the same satellites as TOMS, analysis of larger-scale patterns in stratospheric ozone distribution, and empirical corrections based on field studies. These methods are described in detail at: http://science.larc.nasa.gov/TOR/data.html and in Fishman et al. (2003) and the references cited therein.

- NASA developed the OMI Level 2 tropospheric ozone data set by essentially subtracting MLS stratospheric ozone observations from concurrent OMI total column ozone observations. Ziemke et al. (2006) describe these methods in more detail.

EPA performed the following additional processing steps to convert NASA's data products into an easy-to-understand indicator:

- EPA obtained SBUV data in the form of monthly averages for each of layer of the atmosphere (total: 21 layers) by latitude band (i.e., average ozone levels for each 5-degree band of latitude). For each latitude band, EPA added the ozone levels for NASA's 21 atmospheric layers together to get total column ozone. Next, because each latitude band represents a different amount of surface area of the atmosphere (for example the band near the North Pole from 85°N to 90°N covers a much smaller surface area than the band near the equator from 0° to 5°N), EPA calculated a global average using cosine area weighting. The global average in this indicator only covers the latitude bands between 50°N and 50°S for consistency of satellite coverage. EPA then combined the monthly averages to obtain annual averages.

- EPA obtained TOR and OMI Level 2 data as a grid of monthly average tropospheric ozone levels. Both data sets are divided into grid cells measuring 1 degree latitude by 1.25 degrees longitude and are only available between 50°N and 50°S. EPA calculated global monthly averages for each 1-degree latitude band by averaging over all grid cells in that band, then used cosine area weighting to calculate an average for the entire range from 50°N to 50°S. EPA combined the monthly averages to obtain annual averages.

In Figure 5, the "total column" line comes from the SBUV data set. Because of missing data from mid-1972 through late 1978, EPA elected to start the graph at 1979. From 1979 to present, all years have complete SBUV data.

The "troposphere" line in Figure 5 is based on the TOR data set from 1979 to 2004, the OMI Level 2 data set from 2006 to present, and an average of TOR and OMI Level 2 for 2005. To correct for differences between the two instruments, EPA adjusted all OMI data points upward by 1.799 Dobson units, which is the documented difference during periods of overlap in 2004. This is a standard bootstrapping approach. Data are not shown from 1994 to 1996 because no TOMS instrument was in orbit from late 1994 to early 1996, so it was not possible to calculate annual averages from the TOR data set during these three years. The "stratosphere" line in Figure 5 was calculated by subtracting the "troposphere" series from the "total column" series.

Indicator Development

Figures 1, 2, 3, and 4 were published as part of EPA's 2010 and 2012 climate change indicator reports. EPA added Figure 5 for the 2014 edition to address one of the key limitations of the previous indicator and to reflect the scientific community's growing awareness of the importance of tropospheric ozone as a contributor to climate change.

Scientists measure the amount of ozone in the atmosphere using two complementary methods. In addition to NASA's satellite-based data collection, NOAA operates a set of ground-based sites using devices called Dobson ozone spectrophotometers, which point upward and measure total column ozone on clear days. A set of 10 of these sites constitute the NOAA Ozone Reference Network. Measurements have been collected at some of these sites since the 1920s, and the resulting data are available at: www.esrl.noaa.gov/gmd/ozwv/dobson.

When developing this indicator, EPA chose to focus on satellite measurements because they allow total column ozone to be separated into tropospheric and stratospheric components, which facilitates greater understanding of the complex roles that ozone, ozone-depleting substances, and emissions of ozone precursors play in climate change. In addition to the fact that satellite-based data products were readily available to assess ozone concentrations by layer of the atmosphere, tropospheric ozone is short-lived and not globally mixed, so satellite-based measurements arguably provide more complete coverage of this greenhouse gas than a finite set of ground-based stations. Nonetheless, as described in Section 7, NOAA's ground-based measurements still play an important role in this indicator because they provide independent verification of the trends detected by satellites.

7. Quality Assurance and Quality Control

The data for this indicator have generally been taken from carefully constructed, peer-reviewed studies. Quality assurance and quality control procedures are addressed in the individual studies, which are cited in Section 3. Additional documentation of these procedures can be obtained by consulting with the principal investigators who developed each of the data sets.

NASA selected SBUV data for their official merged ozone data set based on the results of a detailed analysis of instrumental uncertainties (DeLand et al., 2012) and comparisons against independent satellite and ground-based profile observations (Kramarova et al., 2013b; Labow et al., 2013). NASA screened SBUV data using the following rules:

- Data from the SBUV/2 instrument on the NOAA-9 satellite are not included due to multiple instrumental issues (DeLand et al., 2012).

- Only measurements made between 8 AM and 4 PM Equatorial Crossing Time are included in the merged satellite ozone data set, with one exception in 1994-1995, when NOAA 11 data were included to avoid a gap in the data.

- When data from more than one SBUV instrument are available, NASA used a simple average of the data.

- Data were filtered for aerosol contamination after the eruptions of El Chichon (1982) and Mt. Pinatubo (1991).

Satellite data have been validated against ground-based measurements from NOAA's Ozone Reference Network. Figure TD-1 below shows how closely these two complementary data sources track each other over time.

Figure TD-1. Yearly Average Change in Global Total Column Ozone Since 1979

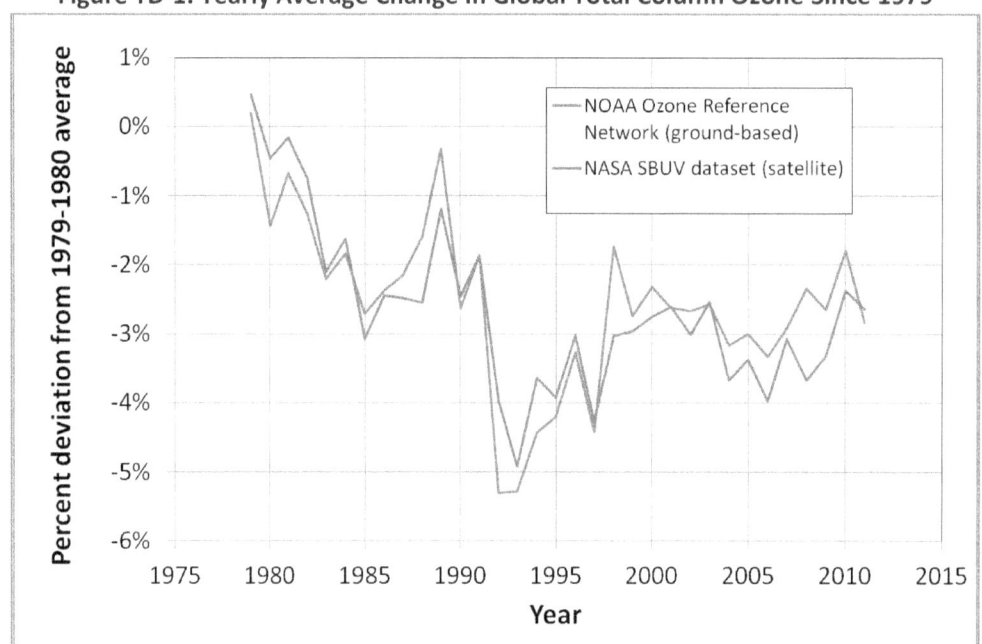

Analysis

8. Comparability Over Time and Space

Data have been collected using a variety of methods over time and space. However, these methodological differences are expected to have little bearing on the overall conclusions for this indicator. The concordance of trends among multiple data sets collected using different program designs provides some assurance that the trends depicted actually represent changes in atmospheric conditions, rather than some artifact of sampling design.

Figures 1, 2, 3, and 4. Global Atmospheric Concentrations of Carbon Dioxide, Methane, Nitrous Oxide, and Selected Halogenated Gases

The gases covered in Figures 1, 2, 3, and 4 are all long-lived GHGs that are relatively evenly distributed globally. Thus, measurements collected at one particular location have been shown to be representative of average concentrations worldwide.

Figure 5. Global Atmospheric Concentrations of Ozone, 1979–2013

Because ozone concentrations vary over time and space, Figure 5 uses data from satellites that cover virtually the entire globe, and the figure shows area-weighted global averages. These satellite data have undergone extensive testing to identify errors and biases by comparing them with independent satellite and ground-based profile observations, including the NOAA reference ozone network (Kramarova et al., 2013b).

9. Data Limitations

Factors that may impact the confidence, application, or conclusions drawn from this indicator are as follows:

1. This indicator does not track water vapor because of its spatial and temporal variability. Human activities have only a small direct impact on water vapor concentrations, but there are indications that increasing global temperatures are leading to increasing levels of atmospheric humidity (Dai et al., 2011).

2. Some radiatively important atmospheric constituents that are substantially affected by human activities (such as black carbon, aerosols, and sulfates) are not included in this indicator because of their spatial and temporal variability.

3. This indicator includes several of the most important halogenated gases, but some others are not shown. Many other halogenated gases are also GHGs, but Figure 4 is limited to a set of common examples that represent most of the major types of these gases.

4. Ice core measurements are not taken in real time, which introduces some error into the date of the sample. Dating accuracy for the ice cores ranges up to plus or minus 20 years (often less), depending on the method used and the time period of the sample. Diffusion of gases from the samples, which would tend to reduce the measured values, could also add a small amount of uncertainty.

5. Factors that could affect satellite-based ozone measurements include orbital drift, instrument differences, and solar zenith angle (the angle of incoming sunlight) at the time of measurement. However, as discussed in Section 10, the data have been filtered and calibrated to account for these factors. For example, Figure 5 has been restricted to the zone between 50°N and 50°S latitude because at higher latitudes the solar zenith angles would introduce greater uncertainty and because the lack of sunlight during winter months creates data gaps.

10. Sources of Uncertainty

Figures 1, 2, 3, and 4. Global Atmospheric Concentrations of Carbon Dioxide, Methane, Nitrous Oxide, and Selected Halogenated Gases

Direct measurements of atmospheric concentrations, which cover approximately the last 50 years, are of a known and high quality. Generally, standard errors and accuracy measurements are computed for the data.

For ice core measurements, uncertainties result from the actual gas measurements as well as the dating of each sample. Uncertainties associated with the measurements are believed to be relatively small, although diffusion of gases from the samples might also add to the measurement uncertainty. Dating accuracy for the ice cores is believed to be within plus or minus 20 years, depending on the method used and the time period of the sample. However, this level of uncertainty is insignificant considering that some ice cores characterize atmospheric conditions for time frames of hundreds of thousands of years. The original scientific publications (see Section 3) provide more detailed information on the estimated uncertainty within the individual data sets.

Visit the Carbon Dioxide Information Analysis Center (CDIAC) website (http://cdiac.esd.ornl.gov/by_new/bysubjec.html#atmospheric) for more information on the accuracy of both direct and ice core measurements.

Overall, the concentration increase in GHGs in the past century is far greater than the estimated uncertainty of the underlying measurement methodologies. It is highly unlikely that the concentration trends depicted in this set of figures are artifacts of uncertainty.

Figure 5. Global Atmospheric Concentrations of Ozone, 1979–2013

NASA has estimated uncertainties for the merged SBUV satellite ozone data set, mostly focusing on errors that might affect trend analysis. Constant offsets and random errors will make no difference in the trend, but smoothing error and instrumental drift can potentially affect trend estimates. The following discussion describes these sources of error, the extent to which they affect the data used for this indicator, and steps taken to minimize the corresponding uncertainty.

- The main source of error in the SBUV data is a smoothing error due to profile variability that the SBUV observing system cannot inherently measure (Bhartia et al., 2012; Kramarova et al., 2013a). NASA's SBUV data set is divided into 21 distinct layers of the atmosphere, and the size of the smoothing error varies depending on the layer. For the layers that make up most of the stratosphere (specifically between 16 hectopascals (hPa) and 1 hPa of pressure in the tropics (20°S to 20°N) and 25 hPa to 1 hPa outside the tropics), the smoothing error for the SBUV monthly mean profiles is approximately 1 percent, indicating that SBUV data are capable of accurately representing ozone changes in this part of the atmosphere. For the SBUV layers that cover the troposphere, the lower stratosphere, and above the stratosphere (air pressure less than 1 hPa), the smoothing errors are larger: up to 8 to 15 percent. The influence of these smoothing errors has been minimized by adding all of the individual SBUV layers together to examine total column ozone.

- Long-term drift can only be estimated through comparison with independent data sources. NASA validated the SBUV merged ozone data set against independent satellite observations and found that drifts are less than 0.3 percent per year and mostly insignificant.

- Several SBUV instruments have been used over time, and each instrument has specific characteristics. NASA estimated the offsets between pairs of SBUV instruments when they overlap (DeLand et al., 2012) and found that mean differences are within 7 percent, with corresponding standard deviations of 1 to 3 percent. The SBUV Version 8.6 algorithm adjusts for these differences based on a precise comparison of radiance measurements during overlap periods.

- Because the SBUV instruments use wavelengths that have high sensitivity to ozone, the total column ozone calculated from this method is estimated to have a 1 to 2 Dobson unit accuracy for solar zenith angles up to 70 degrees—i.e., when the sun is more than 20 degrees above the horizon (Bhartia et al., 2012). Measurements taken when the sun is lower in the sky have less accuracy, which is why the SBUV data in this indicator have been mostly limited to measurements made between 8 AM and 4 PM Equatorial Crossing Time, and one reason why the data have been limited to the area between 55°N and 55°S latitude (82 percent of the Earth's surface area).

Fishman et al. (2003) describe uncertainties in the TOR tropospheric ozone data set. Calculations of global average TOR may vary by up to 5 Dobson units, depending on which release of satellite data is used. For information about uncertainty in the OMI Level 2 tropospheric ozone data set, see Ziemke et al. (2006), which describes in detail how OMI data have been validated against ozonesonde data. Both of these data sets have been limited to the zone between 50°N and 50°S latitude because of the solar angle limitation described above. Based on the considerations, adjustment steps, and validation steps described above, it is unlikely that the patterns depicted in Figure 5 are artifacts of uncertainty.

11. Sources of Variability

Figures 1, 2, 3, and 4. Global Atmospheric Concentrations of Carbon Dioxide, Methane, Nitrous Oxide, and Selected Halogenated Gases

Atmospheric concentrations of the long-lived GHGs vary with both time and space. However, the data presented in this indicator have extraordinary temporal coverage. For carbon dioxide, methane, and nitrous oxide, concentration data span several hundred thousand years; and for the halogenated gases, data span virtually the entire period during which these largely synthetic gases were widely used. While spatial coverage of monitoring stations is more limited, most of the GHGs presented in this indicator are considered to be well-mixed globally, due in large part to their long residence times in the atmosphere.

Figure 5. Global Atmospheric Concentrations of Ozone, 1979–2013

Unlike the other gases described in this indicator, ozone is relatively short-lived in the troposphere, with a typical lifetime of only a few weeks. Concentrations of both tropospheric and stratospheric ozone vary spatially at any given time; for example, Fishman et al. (2003) use the TOR to show noticeably elevated levels of tropospheric ozone over heavily populated and industrialized regions. Fishman et al. (2003) also show seasonal variations. This indicator accounts for both spatial and temporal variations by presenting global annual averages.

12. Statistical/Trend Analysis

This indicator presents a time series of atmospheric concentrations of GHGs. For the long-lived gases, no statistical techniques or analyses have been used to characterize the long-term trends or their statistical significance. For ozone, EPA used ordinary least-squares linear regressions to assess whether changes in ozone levels over time have been statistically significant, which yielded the following results:

- A regression of the total column ozone data from 1979 to 2013 shows a significant decrease of approximately 0.2 Dobson units per year ($p < 0.001$).

- Further analysis of the total column ozone data shows a rapid decline over the first decade and a half of the data record (1979–1994), with insignificant change after that. A regression analysis for 1979–1994 shows a significant decline of about 0.7 Dobson units per year ($p < 0.001$), while the regression for the remainder of the data record (1995–2013) shows an insignificant change ($p = 0.43$).

- A regression of tropospheric ozone from 1979 to 2013 shows a significant increase of 0.02 Dobson units per year ($p = 0.028$).

To conduct a more complete analysis would potentially require consideration of serial correlation and other more complex statistical factors.

References

AGAGE (Advanced Global Atmospheric Gases Experiment). 2011. Data provided to ERG (an EPA contractor) by Ray Wang, Georgia Institute of Technology. November 2011.

AGAGE (Advanced Global Atmospheric Gases Experiment). 2014a. ALE/GAGE/AGAGE data base. Accessed May 2014. http://agage.eas.gatech.edu/data.htm.

AGAGE (Advanced Global Atmospheric Gases Experiment). 2014b. Monthly mean N_2O concentrations for Cape Grim, Australia. Accessed April 8, 2014. http://ds.data.jma.go.jp/gmd/wdcgg/cgi-bin/wdcgg/catalogue.cgi.

Arnold, T., J. Mühle, P.K. Salameh, C.M. Harth, D.J. Ivy, and R.F. Weiss. 2012. Automated measurement of nitrogen trifluoride in ambient air. Analytical Chemistry 84(11):4798–4804.

Arnold, T., C.M. Harth, J. Mühle, A.J. Manning, P.K. Salameh, J. Kim, D.J. Ivy, L.P. Steele, V.V. Petrenko, J.P. Severinghaus, D. Baggenstos, and R.F. Weiss. 2013. Nitrogen trifluoride global emissions estimated from updated atmospheric measurements. PNAS 110(6):2029-2034.

Arnold, T. 2013 update to data originally published in: Arnold, T., C.M. Harth, J. Mühle, A.J. Manning, P.K. Salameh, J. Kim, D.J. Ivy, L.P. Steele, V.V. Petrenko, J.P. Severinghaus, D. Baggenstos, and R.F. Weiss. 2013. Nitrogen trifluoride global emissions estimated from updated atmospheric measurements. P. Natl. Acad. Sci. USA 110(6):2029–2034. Data updated May 2013.

Battle, M., M. Bender, T. Sowers, P. Tans, J. Butler, J. Elkins, J. Ellis, T. Conway, N. Zhang, P. Lang, and A. Clarke. 1996. Atmospheric gas concentrations over the past century measured in air from firn at the South Pole. Nature 383:231–235. ftp://daac.ornl.gov/data/global_climate/global_N_cycle/data/global_N_perturbations.txt.

Bhartia, P.K., R.D. McPeters, L.E. Flynn, S. Taylor, N.A. Kramarova, S. Frith, B. Fisher, and M. DeLand. 2012. Solar Backscatter UV (SBUV) total ozone and profile algorithm. Atmos. Meas. Tech. Discuss. 5:5913–5951.

Chamard, P., L. Ciattaglia, A. di Sarra, and F. Monteleone. 2001. Atmospheric carbon dioxide record from flask measurements at Lampedusa Island. In: Trends: A compendium of data on global change. Oak Ridge, TN: U.S. Department of Energy. Accessed September 14, 2005. http://cdiac.ornl.gov/trends/co2/lampis.html.

Dai, A., J. Wang, P.W. Thorne, D.E. Parker, L. Haimberger, and X.L. Wang. 2011. A new approach to homogenize daily radiosonde humidity data. J. Climate 24(4):965–991. http://journals.ametsoc.org/doi/abs/10.1175/2010JCLI3816.1.

DeLand, M.T., S.L. Taylor, L.K. Huang, and B.L. Fisher. 2012. Calibration of the SBUV version 8.6 ozone data product. Atmos. Meas. Tech. Discuss. 5:5151–5203.

Etheridge, D.M., L.P. Steele, R.L. Langenfelds, R.J. Francey, J.M. Barnola, and V.I. Morgan. 1998. Historical CO_2 records from the Law Dome DE08, DE08-2, and DSS ice cores. In: Trends: A compendium of data on global change. Oak Ridge, TN: U.S. Department of Energy. Accessed September 14, 2005. http://cdiac.ornl.gov/trends/co2/lawdome.html.

Etheridge, D.M., L.P. Steele, R.J. Francey, and R.L. Langenfelds. 2002. Historic CH_4 records from Antarctic and Greenland ice cores, Antarctic firn data, and archived air samples from Cape Grim, Tasmania. In: Trends: A compendium of data on global change. Oak Ridge, TN: U.S. Department of Energy. Accessed September 13, 2005. http://cdiac.ornl.gov/trends/atm_meth/lawdome_meth.html.

Fishman, J., A.E. Wozniak, and J.K. Creilson. 2003. Global distribution of tropospheric ozone from satellite measurements using the empirically corrected tropospheric ozone residual technique: Identification of the regional aspects of air pollution. Atmos. Chem. Phys. 3:893–907.

IPCC (Intergovernmental Panel on Climate Change). 2013. Climate change 2013: The physical science basis. Working Group I contribution to the IPCC Fifth Assessment Report. Cambridge, United Kingdom: Cambridge University Press. www.ipcc.ch/report/ar5/wg1.

Kramarova, N.A., P.K. Bhartia, S.M. Frith, R.D. McPeters, and R.S. Stolarski. 2013a. Interpreting SBUV smoothing errors: An example using the quasi-biennial oscillation. Atmos. Meas. Tech. 6:2089–2099.

Kramarova, N.A., S.M. Frith, P.K. Bhartia, R.D. McPeters, S.L. Taylor, B.L. Fisher, G.J. Labow, and M.T. DeLand. 2013b. Validation of ozone monthly zonal mean profiles obtained from the version 8.6 Solar Backscatter Ultraviolet algorithm. Atmos. Chem. Phys. 13:6887–6905.

Labow, G.J., R.D. McPeters, P.K. Bhartia, and N.A. Kramarova. 2013. A comparison of 40 years of SBUV measurements of column ozone with data from the Dobson/Brewer Network. J. Geophys. Res. Atmos. 118(13):7370–7378.

Loulergue, L., A. Schilt, R. Spahni, V. Masson-Delmotte, T. Blunier, B. Lemieux, J.-M. Barnola, D. Raynaud, T.F. Stocker, and J. Chappellaz. 2008. Orbital and millennial-scale features of atmospheric CH_4 over the past 800,000 years. Nature 453:383–386. www.ncdc.noaa.gov/paleo/pubs/loulergue2008/loulergue2008.html.

Lüthi, D., M. Le Floch, B. Bereiter, T. Blunier, J.-M. Barnola, U. Siegenthaler, D. Raynaud, J. Jouzel, H. Fischer, K. Kawamura, and T.F. Stocker. 2008. High-resolution carbon dioxide concentration record

650,000–800,000 years before present. Nature 453:379–382.
www.ncdc.noaa.gov/paleo/pubs/luethi2008/luethi2008.html.

McPeters, R.D., P.K. Bhartia, D. Haffner, G.J. Labow, and L. Flynn. 2013. The v8.6 SBUV ozone data record: An overview. J. Geophys. Res. Atmos. 118(14):8032–8039.

NASA (National Aeronautics and Space Administration). 2013. Data—TOMS/SBUV TOR data products. Accessed November 2013. http://science.larc.nasa.gov/TOR/data.html.

NASA (National Aeronautics and Space Administration). 2014a. SBUV merged ozone dataset (MOD). Version 8.6. Pre-online release provided by NASA staff, May 2014. http://acdb-ext.gsfc.nasa.gov/Data_services/merged/index.html.

NASA (National Aeronautics and Space Administration). 2014b. Tropospheric ozone data from AURA OMI/MLS. Accessed May 2014. http://acdb-ext.gsfc.nasa.gov/Data_services/cloud_slice/new_data.html.

Neftel, A., H. Friedli, E. Moor, H. Lötscher, H. Oeschger, U. Siegenthaler, and B. Stauffer. 1994. Historical carbon dioxide record from the Siple Station ice core. In: Trends: A compendium of data on global change. Oak Ridge, TN: U.S. Department of Energy. Accessed September 14, 2005. http://cdiac.ornl.gov/trends/co2/siple.html.

NOAA (National Oceanic and Atmospheric Administration). 2013. Halocarbons and Other Atmospheric Trace Species group (HATS). Accessed July 2013. www.esrl.noaa.gov/gmd/hats.

NOAA (National Oceanic and Atmospheric Administration). 2014a. Annual mean carbon dioxide concentrations for Mauna Loa, Hawaii. Accessed April 7, 2014. ftp://ftp.cmdl.noaa.gov/products/trends/co2/co2_annmean_mlo.txt.

NOAA (National Oceanic and Atmospheric Administration). 2014b. Data provided to ERG (an EPA contractor) by Stephen Montzka, NOAA. April 2014.

NOAA (National Oceanic and Atmospheric Administration). 2014c. Monthly mean carbon dioxide concentrations for Barrow, Alaska; Cape Matatula, American Samoa; and the South Pole. Accessed April 7, 2014. ftp://ftp.cmdl.noaa.gov/data/trace_gases/co2/in-situ.

NOAA (National Oceanic and Atmospheric Administration). 2014d. Monthly mean CH_4 concentrations for Cape Grim, Australia. Accessed April 8, 2014. ftp://ftp.cmdl.noaa.gov/data/trace_gases/ch4/flask/surface/ch4_cgo_surface-flask_1_ccgg_month.txt.

NOAA (National Oceanic and Atmospheric Administration). 2014e. Monthly mean CH_4 concentrations for Mauna Loa, Hawaii. Accessed April 8, 2014. ftp://ftp.cmdl.noaa.gov/data/trace_gases/ch4/in-situ/surface/mlo/ch4_mlo_surface-insitu_1_ccgg_month.txt.

NOAA (National Oceanic and Atmospheric Administration). 2014f. Monthly mean N_2O concentrations for Barrow, Alaska; Mauna Loa, Hawaii; and the South Pole. Accessed April 8, 2014. www.esrl.noaa.gov/gmd/hats/insitu/cats/cats_conc.html.

Schilt, A., M. Baumgartner, T. Blunier, J. Schwander, R. Spahni, H. Fischer, and T.F. Stocker. 2010. Glacial-interglacial and millennial scale variations in the atmospheric nitrous oxide concentration during the last 800,000 years. Quaternary Sci. Rev. 29:182–192. ftp://ftp.ncdc.noaa.gov/pub/data/paleo/icecore/antarctica/epica_domec/edc-n2o-2010-800k.txt.

Steele, L.P., P.B. Krummel, and R.L. Langenfelds. 2002. Atmospheric methane record from Shetland Islands, Scotland (October 2002 version). In: Trends: A compendium of data on global change. Oak Ridge, TN: U.S. Department of Energy. Accessed September 13, 2005. http://cdiac.esd.ornl.gov/trends/atm_meth/csiro/csiro-shetlandch4.html.

Steele, L.P., P.B. Krummel, and R.L. Langenfelds. 2007. Atmospheric CO_2 concentrations (ppmv) derived from flask air samples collected at Cape Grim, Australia, and Shetland Islands, Scotland. Commonwealth Scientific and Industrial Research Organisation. Accessed January 20, 2009. http://cdiac.esd.ornl.gov/ftp/trends/co2/csiro.

Weiss, R.F., J. Mühle, P.K. Salameh, and C.M. Harth. 2008. Nitrogen trifluoride in the global atmosphere. Geophys. Res. Lett. 35:L20821.

Ziemke, J.R., S. Chandra, B.N. Duncan, L. Froidevaux, P.K. Bhartia, P.F. Levelt, and J.W. Waters. 2006. Tropospheric ozone determined from Aura OMI and MLS: Evaluation of measurements and comparison with the Global Modeling Initiative's Chemical Transport Model. J. Geophys. Res. 111:D19303.

Climate Forcing

Identification

1. Indicator Description

This indicator measures the levels of greenhouse gases (GHGs) in the atmosphere based on their ability to cause changes in the Earth's climate. This indicator is highly relevant to climate change because greenhouse gases from human activities are the primary driver of observed climate change since the mid-20th century (IPCC, 2013). Components of this indicator include:

- Radiative forcing associated with long-lived GHGs as measured by the Annual Greenhouse Gas Index from 1979 to 2013 (Figure 1).

- A reference figure showing estimates of total radiative forcing associated with a variety of human activities since the year 1750 (Figure 2).

2. Revision History

April 2010:   Indicator posted.
December 2011:   Indicator updated with data through 2010.
October 2012:   Indicator updated with data through 2011.
December 2013:   Indicator updated with data through 2012; added Figure 2 to provide longer-term context and cover other climate forcers.
May 2014:   Updated Figure 1 with data through 2013.

Data Sources

3. Data Sources

GHG concentrations for Figure 1 are measured by a cooperative global network of monitoring stations overseen by the National Oceanic and Atmospheric Administration's (NOAA's) Earth System Research Laboratory (ESRL). The figure uses measurements of 20 GHGs.

Estimates of total radiative forcing in Figure 2 were provided by the Intergovernmental Panel on Climate Change (IPCC) and published in the IPCC's Fifth Assessment Report (IPCC, 2013).

4. Data Availability

Figure 1. Radiative Forcing Caused by Major Long-Lived Greenhouse Gases, 1979–2013

Figure 1 is based on NOAA's Annual Greenhouse Gas Index (AGGI). Annual values of the AGGI (total and broken down by gas) are posted online at: www.esrl.noaa.gov/gmd/aggi/, along with definitions and descriptions of the data. EPA obtained data from NOAA's public website.

The AGGI is based on data from monitoring stations around the world. Most of these data were collected as part of the NOAA/ESRL cooperative monitoring network. Data files from these cooperative stations are available online at: www.esrl.noaa.gov/gmd/dv/ftpdata.html. Users can obtain station metadata by navigating to: www.esrl.noaa.gov/gmd/dv/site/, viewing a list of stations, and then selecting a station of interest.

Methane data prior to 1983 are annual averages from Etheridge et al. (1998). Users can download data from this study at: http://cdiac.ornl.gov/trends/atm_meth/lawdome_meth.html.

Figure 2. Radiative Forcing Caused by Human Activities Since 1750

Figure 2 is adapted from a figure in IPCC's Fifth Assessment Report (IPCC, 2013). The original figure is available at: www.climatechange2013.org/report/reports-graphic. Underlying data came from a broad assessment of the best available scientific literature. Specific sources are cited in IPCC (2013).

Methodology

5. Data Collection

Figure 1. Radiative Forcing Caused by Major Long-Lived Greenhouse Gases, 1979–2013

The AGGI is based on measurements of the concentrations of various long-lived GHGs in ambient air. These measurements have been collected following consistent high-precision techniques that have been documented in peer-reviewed literature.

The indicator uses measurements of five major GHGs and 15 other GHGs. The five major GHGs for this indicator are carbon dioxide (CO_2), methane (CH_4), nitrous oxide (N_2O), and two chlorofluorocarbons, CFC-11 and CFC-12. According to NOAA, these five GHGs account for approximately 96 percent of the increase in direct radiative forcing by long-lived GHGs since 1750. The other 15 gases are CFC-113, carbon tetrachloride (CCl_4), methyl chloroform (CH_3CCl_3), HCFC-22, HCFC-141b, HCFC-142b, HFC-23, HFC-125, HFC-134a, HFC-143a, HFC-152a, sulfur hexafluoride (SF_6), halon-1211, halon-1301, and halon-2402.

Monitoring stations in NOAA's ESRL network collect air samples at approximately 80 global clean air sites, although not all sites monitor for all the gases of interest. Monitoring sites include fixed stations on land as well as measurements at 5-degree latitude intervals along specific ship routes in the oceans. Monitoring stations collect data at least weekly. These weekly measurements can be averaged to arrive at an accurate representation of annual concentrations.

For a map of monitoring sites in the NOAA/ESRL cooperative network, see: www.esrl.noaa.gov/gmd/aggi. For more information about the global monitoring network and a link to an interactive map, see NOAA's website at: www.esrl.noaa.gov/gmd/dv/site.

Figure 2. Radiative Forcing Caused by Human Activities Since 1750

The broader reference figure presents the best available estimates of radiative forcing from 1750 through 2011, based on the IPCC's complete assessment of the scientific literature. Thus, this part of the indicator reflects a large number of studies and monitoring programs.

6. Indicator Derivation

Figure 1. Radiative Forcing Caused by Major Long-Lived Greenhouse Gases, 1979–2013

From weekly station measurements, NOAA calculated a global average concentration of each gas using a smoothed north-south latitude profile in sine latitude space, which essentially means that the global average accounts for the portion of the Earth's surface area contained within each latitude band. NOAA averaged these weekly global values over the course of the year to determine an annual average concentration of each gas. Pre-1983 methane measurements came from stations outside the NOAA/ESRL network; these data were adjusted to NOAA's calibration scale before being incorporated into the indicator.

Next, NOAA transformed gas concentrations into an index based on radiative forcing. These calculations account for the fact that different gases have different abilities to alter the Earth's energy balance. NOAA determined the total radiative forcing of the GHGs by applying radiative forcing factors that have been scientifically established for each gas based on its global warming potential and its atmospheric lifetime. These values and equations were published in the Intergovernmental Panel on Climate Change's (IPCC's) Third Assessment Report (IPCC, 2001). In order to keep the index as accurate as possible, NOAA's radiative forcing calculations considered only direct forcing, not additional model-dependent feedbacks, such as those due to water vapor and ozone depletion.

NOAA compared present-day concentrations with those circa 1750 (i.e., before the start of the Industrial Revolution), and this indicator shows only the radiative forcing associated with the *increase* in concentrations since 1750. In this regard, the indicator focuses only on the additional radiative forcing that has resulted from human-influenced emissions of GHGs.

Figure 1 shows radiative forcing from the selected GHGs in units of watts per square meter (W/m^2). This forcing value is calculated at the tropopause, which is the boundary between the troposphere and the stratosphere. Thus, the square meter term refers to the surface area of the sphere that contains the Earth and its lower atmosphere (the troposphere). The watts term refers to the rate of energy transfer.

The data provided to EPA by NOAA also describe radiative forcing in terms of the AGGI. This unitless index is formally defined as the ratio of radiative forcing in a given year compared with a base year of 1990, which was chosen because 1990 is the baseline year for the Kyoto Protocol. Thus, 1990 is set to a total AGGI value of 1. An AGGI scale appears on the right side of Figure 1.

NOAA's monitoring network did not provide sufficient data prior to 1979, and no attempt has been made to project the indicator backward before that start date. No attempt has been made to project trends forward into the future, either.

This indicator can be reconstructed from publicly available information. NOAA's website (www.esrl.noaa.gov/gmd/aggi) provides a complete explanation of how to construct the AGGI from the available concentration data, including references to the equations used to determine each gas's contribution to radiative forcing. See Hofmann et al. (2006a) and Hofmann et al. (2006b) for more information about the AGGI and how it was constructed. See Dlugokencky et al. (2005) for information on steps that were taken to adjust pre-1983 methane data to NOAA's calibration scale.

Figure 2. Radiative Forcing Caused by Human Activities Since 1750

EPA used the data in Figure 2 exactly as they were provided by IPCC. EPA modified the original figure text in a few ways to make it easier for readers to understand, such as by explaining that albedo refers to the reflectivity of a surface.

Indicator Development

Figure 1 was published as part of EPA's 2010 and 2012 climate change indicator reports. EPA added Figure 2 for the 2014 edition to address some of the key limitations of the previous version of the indicator, and to reflect the scientific community's growing awareness of the importance of tropospheric ozone and black carbon as contributors to climate change.

7.   Quality Assurance and Quality Control

Figure 1. Radiative Forcing Caused by Major Long-Lived Greenhouse Gases, 1979–2013

The online documentation for the AGGI does not explicitly discuss quality assurance and quality control (QA/QC) procedures. NOAA's analysis has been peer-reviewed and published in the scientific literature, however (see Hofmann et al., 2006a and 2006b), and users should have confidence in the quality of the data.

Figure 2. Radiative Forcing Caused by Human Activities Since 1750

IPCC (2013) describes the careful review that went into selecting sources for the Fifth Assessment Report. The original peer-reviewed studies cited therein provide more details about specific QA/QC protocols.

Analysis

8.   Comparability Over Time and Space

Figure 1. Radiative Forcing Caused by Major Long-Lived Greenhouse Gases, 1979–2013

With the exception of pre-1983 methane measurements, all data were collected through the NOAA/ESRL global monitoring network with consistent techniques over time and space. Pre-1983 methane measurements came from stations outside the NOAA/ESRL network; these data were adjusted to NOAA's calibration scale before being incorporated into the indicator.

The data for this indicator have been spatially averaged to ensure that the final value for each year accounts for all of the original measurements to the appropriate degree. Results are considered to be globally representative, which is an appropriate assumption because the gases covered by this indicator have long residence times in the atmosphere and are considered to be well-mixed. Although there are minor variations among sampling locations, the overwhelming consistency among sampling locations indicates that extrapolation from these locations to the global atmosphere is reliable.

Figure 2. Radiative Forcing Caused by Human Activities Since 1750

When aggregating data for Figure 2, IPCC selected the best available sources of globally representative information. Total radiative forcing has been aggregated over time from 1750 to 2011.

9. Data Limitations

Factors that may impact the confidence, application, or conclusions drawn from this indicator are as follows:

1. The AGGI and its underlying analysis do not provide a complete picture of radiative forcing from the major GHGs because they do not consider indirect forcing due to water vapor, ozone depletion, and other factors. These mechanisms have been excluded because quantifying them would require models that would add substantial uncertainty to the indicator.

2. The AGGI also does not include radiative forcing due to shorter-lived GHGs and other radiatively important atmospheric constituents, such as black carbon, aerosols, and sulfates. Reflective aerosol particles in the atmosphere can reduce climate forcing, for example, while tropospheric ozone can increase it. These spatially heterogeneous, short-lived climate forcing agents have uncertain global magnitudes and thus are excluded from NOAA's index to maintain accuracy. These factors have been addressed at a broader scale in Figure 2 for reference.

10. Sources of Uncertainty

Figure 1. Radiative Forcing Caused by Major Long-Lived Greenhouse Gases, 1979–2013

This indicator is based on direct measurements of atmospheric concentrations of GHGs. These measurements are of a known and high quality, collected by a well-established monitoring network. NOAA's AGGI website does not present explicit uncertainty values for either the AGGI or the underlying data, but exact uncertainty estimates can be obtained by contacting NOAA.

The empirical expressions used for radiative forcing are derived from atmospheric radiative transfer models and generally have an uncertainty of about 10 percent. The uncertainties in the global average concentrations of the long-lived GHGs are much smaller, according to the AGGI website documentation at: www.esrl.noaa.gov/gmd/aggi.

Uncertainty is expected to have little bearing on the conclusions for several reasons. First, the indicator is based entirely on measurements that have low uncertainty. Second, the increase in GHG radiative forcing over recent years is far greater than the estimated uncertainty of underlying measurement methodologies, and it is also greater than the estimated 10 percent uncertainty in the radiative forcing equations. Thus, it is highly unlikely that the trends depicted in this indicator are somehow an artifact of uncertainties in the sampling and analytical methods.

Figure 2. Radiative Forcing Caused by Human Activities Since 1750

The colored bars in Figure 2 show IPCC's best central estimates of radiative forcing associated with various human activities, based on the range of values published in the scientific literature. Figure 2 also shows error bars that reflect the likely range of values for each estimate. The original version of IPCC's figure at: www.climatechange2013.org/report/reports-graphic provides the numbers associated with these ranges, and it also classifies the level of scientific understanding for each category as either high, medium, or low. For example, the scientific community's level of understanding of long-lived GHGs is considered to be high to very high, understanding of aerosols such as black carbon is considered high, understanding of short-lived gases is considered medium, and understanding of cloud adjustment due to aerosols is considered low. Overall, IPCC estimates a net radiative forcing associated with human activities of +2.29 W/m^2 since 1750, with a range of +1.13 to +3.33 W/m^2 (IPCC, 2013).

11. Sources of Variability

Collecting data from different locations could lead to some variability, but this variability is expected to have little bearing on the conclusions. Scientists have found general agreement in trends among multiple data sets collected at different locations using different program designs, providing some assurance that the trends depicted actually represent atmospheric conditions, rather than some artifact of sampling design.

12. Statistical/Trend Analysis

The increase in GHG radiative forcing over recent years is far greater than the estimated uncertainty of underlying measurement methodologies, and it is also greater than the estimated 10 percent uncertainty in the radiative forcing equations. Thus, it is highly likely that the trends depicted in this indicator accurately represent changes in the Earth's atmosphere.

References

Dlugokencky, E.J., R.C. Myers, P.M. Lang, K.A. Masarie, A.M. Crotwell, K.W. Thoning, B.D. Hall, J.W. Elkins, and L.P Steele. 2005. Conversion of NOAA atmospheric dry air CH_4 mole fractions to a gravimetrically-prepared standard scale. J. Geophys. Res. 110:D18306.

Etheridge, D.M., L.P. Steele, R.J. Francey, and R.L. Langenfelds. 1998. Atmospheric methane between 1000 A.D. and present: Evidence of anthropogenic emissions and climate variability. J. Geophys. Res. 103:15,979–15,993.

Hofmann, D.J., J.H. Butler, E.J. Dlugokencky, J.W. Elkins, K. Masarie, S.A. Montzka, and P. Tans. 2006a. The role of carbon dioxide in climate forcing from 1979–2004: Introduction of the Annual Greenhouse Gas Index. Tellus B. 58B:614–619.

Hofmann, D.J., J.H. Butler, T.J. Conway, E.J. Dlugokencky, J.W. Elkins, K. Masarie, S.A. Montzka, R.C. Schnell, and P. Tans. 2006b. Tracking climate forcing: The Annual Greenhouse Gas Index. Eos T. Am. Geophys. Un. 87:509–511.

IPCC (Intergovernmental Panel on Climate Change). 2001. Climate change 2001: The scientific basis. Working Group I contribution to the IPCC Third Assessment Report. Cambridge, United Kingdom: Cambridge University Press. www.ipcc.ch/ipccreports/tar/wg1/index.htm.

IPCC (Intergovernmental Panel on Climate Change). 2013. Climate change 2013: The physical science basis. Working Group I contribution to the IPCC Fifth Assessment Report. Cambridge, United Kingdom: Cambridge University Press. www.ipcc.ch/report/ar5/wg1.

U.S. and Global Temperature

Identification

1. Indicator Description

This indicator describes changes in average air temperature for the United States and the world from 1901 to 2013. In this indicator, temperature data are presented as trends in anomalies. Air temperature is an important component of climate, and changes in temperature can have wide-ranging direct and indirect effects on the environment and society.

Components of this indicator include:

- Changes in temperature in the contiguous 48 states over time (Figure 1).
- Changes in temperature worldwide over time (Figure 2).
- A map showing rates of temperature change across the United States (Figure 3).

2. Revision History

April 2010:	Indicator posted.
December 2011:	Updated with data through 2010.
May 2012:	Updated with data through 2011.
August 2013:	Updated indicator on EPA's website with data through 2012.
May 2014:	Updated Figures 1 and 2 with data through 2013.

Data Sources

3. Data Sources

This indicator is based on temperature anomaly data provided by the National Oceanic and Atmospheric Administration's (NOAA's) National Climatic Data Center (NCDC).

4. Data Availability

The long-term surface time series in Figures 1, 2, and 3 were provided to EPA by NOAA's NCDC. NCDC calculated these time series based on monthly values from a set of NCDC-maintained data sets: the *n*ClimDiv data set, the U.S. Historical Climatology Network (USHCN) Version 2.5, the Global Historical Climatology Network–Monthly (GHCN-M) Version 3.2.0 (for global time series, as well as underlying data that feed into *n*ClimDiv), and GHCN-Daily Version 3.0.2 (for Alaska and Hawaii maps). These data sets can be accessed online. To supplement Figures 1 and 2, EPA obtained satellite-based measurements from NCDC's public website.

Contiguous 48 States and Global Time Series (Surface)

Surface time series for the contiguous 48 states (Figure 1) and the world (Figure 2) were obtained from NCDC's "Climate at a Glance" Web interface (www.ncdc.noaa.gov/cag), which draws data from the *n*ClimDiv data set for the contiguous 48 states and from GHCN for global analyses. The *n*ClimDiv product incorporates data from GHCN-Daily and is updated once a month. For access to *n*ClimDiv data and documentation, see: www.ncdc.noaa.gov/monitoring-references/maps/us-climate-divisions.php.

Contiguous 48 States Map (Surface)

Underlying temperature data for the map of the contiguous 48 states (Figure 3) come from the USHCN. Currently, the data are distributed by NCDC on various computer media (e.g., anonymous FTP sites), with no confidentiality issues limiting accessibility. Users can link to the data online at: www.ncdc.noaa.gov/oa/climate/research/ushcn/#access. Appropriate metadata and "readme" files are appended to the data. For example, see: ftp://ftp.ncdc.noaa.gov/pub/data/ushcn/v2/monthly/readme.txt.

Alaska and Hawaii Maps (Surface)

Because the USHCN is limited to the contiguous 48 states, the Alaska and Hawaii portions of the map (Figure 3) are based on data from the GHCN. GHCN temperature data can be obtained from NCDC over the Web or via anonymous FTP. This indicator is specifically based on a combined global land-sea temperature data set that can be obtained from: www.ncdc.noaa.gov/ghcnm/v3.php. There are no known confidentiality issues that limit access to the data set, and the data are accompanied by metadata.

Satellite Data

EPA obtained the satellite trends from NCDC's public website at: www.ncdc.noaa.gov/oa/climate/research/msu.html.

Methodology

5. Data Collection

This indicator is based on temperature measurements. The global portion of this indicator presents temperatures measured over land and sea, while the portion devoted to the contiguous 48 states shows temperatures measured over land only.

Surface data for this indicator were compiled from thousands of weather stations throughout the United States and worldwide using standard meteorological instruments. All of the networks of stations cited here are overseen by NOAA, and their methods of site selection and quality control have been extensively peer reviewed. As such, they represent the most complete long-term instrumental data sets for analyzing recent climate trends. More information on these networks can be found below.

USHCN Surface Data

USHCN Version 2.5 contains monthly averaged maximum, minimum, and mean temperature data from approximately 1,200 stations within the contiguous 48 states. The period of record varies for each station but generally includes most of the 20th century. One of the objectives in establishing the USHCN was to detect secular changes in regional rather than local climate. Therefore, stations included in the network are only those believed to not be influenced to any substantial degree by artificial changes of local environments. Some of the stations in the USHCN are first-order weather stations, but the majority are selected from approximately 5,000 cooperative weather stations in the United States. To be included in the USHCN, a station has to meet certain criteria for record longevity, data availability (percentage of available values), spatial coverage, and consistency of location (i.e., experiencing few station changes). An additional criterion, which sometimes compromised the preceding criteria, was the desire to have a uniform distribution of stations across the United States. Included with the data set are metadata files that contain information about station moves, instrumentation, observation times, and elevation. NOAA's website provides more information about USHCN data collection: www.ncdc.noaa.gov/oa/climate/research/ushcn.

GHCN Surface Data

Because the USHCN is limited to the contiguous 48 states, the Alaska and Hawaii portions of the map are based on data from the GHCN, which contains daily and monthly climate data from weather stations worldwide—including stations within the contiguous 48 states. Monthly mean temperature data are available for 7,280 stations, with homogeneity-adjusted data available for a subset (5,206 mean temperature stations). Data were obtained from many types of stations. For the global component of this indicator, the GHCN land-based data were merged with an additional set of long-term sea surface temperature data. This merged product is called the extended reconstructed sea surface temperature (ERSST) data set, Version #3b (Smith et al., 2008).

NCDC has published documentation for the GHCN. For more information, including data sources, methods, and recent improvements, see: www.ncdc.noaa.gov/ghcnm/v3.php and the sources listed therein. Additional background on the merged land-sea temperature data set can be found at: www.ncdc.noaa.gov/cmb-faq/anomalies.html.

nClimDiv Surface Data

The new *n*ClimDiv divisional data set incorporates data from GHCN-Daily stations in the contiguous 48 states. This data set includes stations that were previously part of the USHCN, as well as additional stations that were able to be added to *n*ClimDiv as a result of quality-control adjustments and digitization of paper records. Altogether, *n*ClimDiv incorporates data from more than 10,000 stations.

In addition to incorporating more stations, the *n*ClimDiv data set differs from the USHCN because it incorporates a grid-based computational approach known as climatologically-aided interpolation (Willmott and Robeson, 1995), which helps to address topographic variability. Data from individual stations are combined in a grid that covers the entire contiguous 48 states with 5-kilometer resolution. These improvements have led to a new data set that maintains the strengths of its predecessor data sets while providing more robust estimates of area averages and long-term trends. The *n*ClimDiv data set is NOAA's official temperature data set for the contiguous 48 states, replacing USHCN.

To learn more about *n*ClimDiv, see: www.ncdc.noaa.gov/news/ncdc-introduces-national-temperature-index-page and: www.ncdc.noaa.gov/monitoring-references/maps/us-climate-divisions.php.

Satellite Data

In Figures 1 and 2, surface measurements have been supplemented with satellite-based measurements for the period from 1979 to 2012. These satellite data were collected by NOAA's polar-orbiting satellites, which take measurements across the entire Earth. Satellites equipped with the necessary measuring equipment have orbited the Earth continuously since 1978, but 1979 was the first year with complete data. This indicator uses measurements that represent the lower troposphere, which is defined here as the layer of the atmosphere extending from the Earth's surface to an altitude of about 8 kilometers.

NOAA's satellites use the Microwave Sounding Unit (MSU) to measure the intensity of microwave radiation given off by various layers of the Earth's atmosphere. The intensity of radiation is proportional to temperature, which can therefore be determined through correlations and calculations. NOAA uses different MSU channels to characterize different parts of the atmosphere. Since 1998, NOAA has used the Advanced MSU, a newer version of the instrument.

For more information about the methods used to collect satellite measurements, see: www.ncdc.noaa.gov/oa/climate/research/msu.html and the references cited therein.

6.   Indicator Derivation

Surface Data

NOAA calculated monthly temperature means for each site. In populating the USHCN, GHCN, and *n*ClimDiv, NOAA adjusted the data to remove biases introduced by differences in the time of observation. NOAA also employed a homogenization algorithm to identify and correct for substantial shifts in local-scale data that might reflect changes in instrumentation, station moves, or urbanization effects. These adjustments were performed according to published, peer-reviewed methods. For more information on these quality assurance and error correction procedures, see Section 7.

In this indicator, temperature data are presented as trends in anomalies. An anomaly is the difference between an observed value and the corresponding value from a baseline period. This indicator uses a baseline period of 1901 to 2000. The choice of baseline period *will not* affect the shape or the statistical significance of the overall trend in anomalies. For temperature (absolute anomalies), it only moves the trend up or down on the graph in relation to the point defined as "zero."
To generate the temperature time series, NOAA converted measurements into monthly anomalies in degrees Fahrenheit. The monthly anomalies then were averaged to determine an annual temperature anomaly for each year.

To achieve uniform spatial coverage (i.e., not biased toward areas with a higher concentration of measuring stations), NOAA averaged anomalies within grid cells on the map to create "gridded" data sets. The graph for the contiguous 48 states (Figure 1) is based on the *n*ClimDiv gridded data set, which reflects a high-resolution (5-kilometer) grid. The map (Figure 3) is based on an analysis using grid cells that measure 2.5 degrees latitude by 3.5 degrees longitude. The global graph (Figure 2) comes from an analysis of grid cells measuring 5 degrees by 5 degrees. These particular grid sizes have been determined to be optimal for analyzing USHCN and GHCN climate data. See: www.ncdc.noaa.gov/oa/climate/research/ushcn/gridbox.html for more information.

Figures 1 and 2 show trends from 1901 to 2013, based on NOAA's gridded data sets. Although earlier data are available for some stations, 1901 was selected as a consistent starting point.

The map in Figure 3 shows long-term rates of change in temperature over the United States for the period 1901–2012, except for Alaska and Hawaii, for which widespread and reliable data collection did not begin until 1918 and 1905, respectively. A regression was performed on the annual anomalies for each grid cell. Trends were calculated only in those grid cells for which data were available for at least 66 percent of the years during the full period of record. The slope of each trend (rate of temperature change per year) was calculated from the annual time series by ordinary least-squares regression and then multiplied by 100 to obtain a rate per century. No attempt has been made to portray data beyond the time and space in which measurements were made.

NOAA is continually refining historical data points in the USHCN and GHCN, often as a result of improved methods to reduce bias and exclude erroneous measurements. These improvements frequently result in the designation of new versions of the USHCN and GHCN. As EPA updates this indicator to reflect these upgrades, slight changes to some historical data points may become apparent.

Satellite Data

NOAA's satellites measure microwave radiation at various frequencies, which must be converted to temperature and adjusted for time-dependent biases using a set of algorithms. Various experts recommend slightly different algorithms. Accordingly, Figure 1 and Figure 2 show globally averaged trends that have been calculated by two different organizations: the Global Hydrology and Climate Center at the University of Alabama in Huntsville (UAH) and Remote Sensing Systems (RSS). For more information about the methods used to convert satellite measurements to temperature readings for various layers of the atmosphere, see: www.ncdc.noaa.gov/oa/climate/research/msu.html and the references cited therein. Both the UAH and RSS data sets are based on updated versions of analyses that have been published in the scientific literature. For example, see Christy et al. (2000, 2003), Mears et al. (2003), and Schabel et al. (2002).

NOAA provided data in the form of monthly anomalies. EPA calculated annual anomalies, then shifted the entire curves vertically in order to display the anomalies side-by-side with surface anomalies. Shifting the curves vertically does not change the shape or magnitude of the trends; it simply results in a new baseline. No attempt has been made to portray satellite-based data beyond the time and space in which measurements were made. The satellite data in Figure 1 are restricted to the atmosphere above the contiguous 48 states.

Indicator Development

Previous versions of this indicator were based entirely on the USHCN and GHCN. NCDC launched nClimDiv in early 2014 as a successor to the USHCN and other products. To learn more about this ongoing transition, see: www.ncdc.noaa.gov/news/transitioning-gridded-climate-divisional-dataset. NCDC's initial release of the *n*ClimDiv data set and the corresponding "Climate at a Glance" Web interface in early 2014 made it possible for EPA to update Figures 1 and 2 with newer data through 2013. However, the large-grid analysis in Figure 3 could not be readily updated with the new data at that time. Thus, Figure 3 continues to show data through 2012, based on the USHCN and GHCN. EPA is working to develop a revised map analysis for future editions of this indicator.

7. Quality Assurance and Quality Control

NCDC's databases have undergone extensive quality assurance procedures to identify errors and biases in the data and either remove these stations from the time series or apply correction factors.

USHCN Surface Data

Quality control procedures for the USHCN are summarized at: www.ncdc.noaa.gov/oa/climate/research/ushcn. Homogeneity testing and data correction methods are described in numerous peer-reviewed scientific papers by NOAA's NCDC. A series of data corrections was developed to specifically address potential problems in trend estimation of the rates of warming or cooling in USHCN Version 2.5. They include:

- Removal of duplicate records.
- Procedures to deal with missing data.
- Adjusting for changes in observing practices, such as changes in observation time.
- Testing and correcting for artificial discontinuities in a local station record, which might reflect station relocation, instrumentation changes, or urbanization (e.g., heat island effects).

GHCN Surface Data

QA/QC procedures for GHCN temperature data are described in detail in Peterson et al. (1998) and Menne and Williams (2009), and at: www.ncdc.noaa.gov/ghcnm. GHCN data undergo rigorous QA reviews, which include pre-processing checks on source data; removal of duplicates, isolated values, and suspicious streaks; time series checks to identify spurious changes in the mean and variance via pairwise comparisons; spatial comparisons to verify the accuracy of the climatological mean and the seasonal cycle; and neighbor checks to identify outliers from both a serial and a spatial perspective.

nClimDiv Surface Data

The new *n*ClimDiv data set follows the USHCN's methods to detect and correct station biases brought on by changes to the station network over time. The transition to a grid-based calculation did not significantly change national averages and totals, but it has led to improved historical temperature values in certain regions, particularly regions with extensive topography above the average station elevation—topography that is now being more thoroughly accounted for. An assessment of the major impacts of the transition to *n*ClimDiv can be found at: ftp://ftp.ncdc.noaa.gov/pub/data/cmb/GrDD-Transition.pdf.

Satellite Data

NOAA follows documented procedures for QA/QC of data from the MSU satellite instruments. For example, see NOAA's discussion of MSU calibration at: www.star.nesdis.noaa.gov/smcd/emb/mscat/algorithm.php.

Analysis

8. Comparability Over Time and Space

Both the USHCN and the GHCN have undergone extensive testing to identify errors and biases in the data and either remove these stations from the time series or apply scientifically appropriate correction factors to improve the utility of the data. In particular, these corrections address changes in the time-of-day of observation, advances in instrumentation, and station location changes.

USHCN Surface Data

Homogeneity testing and data correction methods are described in more than a dozen peer-reviewed scientific papers by NCDC. Data corrections were developed to specifically address potential problems in trend estimation of the rates of warming or cooling in the USHCN (see Section 7 for documentation). Balling and Idso (2002) compared the USHCN data with several surface and upper-air data sets and showed that the effects of the various USHCN adjustments produce a significantly more positive, and likely spurious, trend in the USHCN data. However, Balling and Idso (2002) drew conclusions based on an analysis that is no longer valid, as it relied on the UAH satellite temperature data set before corrections identified by Karl et al. (2006) were applied to the satellite record. These corrections have been accepted by all the researchers involved, including those at UAH, and they increased the temperature trend in the satellite data set, eliminating many of the discrepancies with the surface temperature data set. Additionally, even before these corrections were identified, Vose et al. (2003) found that "the time of observation bias adjustments in HCN appear to be robust," contrary to the assertions of Balling and Idso that these adjustments were biased. Vose et al. (2003) found that USHCN station history information is reasonably complete and that the bias adjustment models have low residual errors.

Further analysis by Menne et al. (2009) suggests that:

> ...the collective impact of changes in observation practice at USHCN stations is systematic and of the same order of magnitude as the background climate signal. For this reason, bias adjustments are essential to reducing the uncertainty in U.S. climate trends. The largest biases in the HCN are shown to be associated with changes to the time of observation and with the widespread changeover from liquid-in-glass thermometers to the maximum minimum temperature sensor (MMTS). With respect to [USHCN] Version 1, Version 2 trends in maximum temperatures are similar while minimum temperature trends are somewhat smaller because of an apparent overcorrection in Version 1 for the MMTS instrument change, and because of the systematic impact of undocumented station changes, which were not addressed [in] Version 1.

USHCN Version 2 represents an improvement in this regard. USHCN Version 2.5 further refines these improvements as described at: www.ncdc.noaa.gov/oa/climate/research/ushcn.

Some observers have expressed concerns about other aspects of station location and technology. For example, Watts (2009) expresses concern that many U.S. weather stations are sited near artificial heat sources such as buildings and paved areas, potentially biasing temperature trends over time. In response to these concerns, NOAA analyzed trends for a subset of stations that Watts had determined to be "good or best," and found the temperature trend over time to be very similar to the trend across the full set of USHCN stations (www.ncdc.noaa.gov/oa/about/response-v2.pdf). NOAA's Climate Reference Network (www.ncdc.noaa.gov/crn), a set of optimally-sited stations completed in 2008, can be used to test the accuracy of recent trends. While it is true that many other stations are not optimally located, NOAA's findings support the results of an earlier analysis by Peterson (2006), who found no significant bias in long-term trends associated with station siting once NOAA's homogeneity adjustments were applied. An independent analysis by the Berkeley Earth Surface Temperature (BEST) project (http://berkeleyearth.org/summary-of-findings) used more stations and a different statistical methodology, yet found similar results.

GHCN Surface Data

The GHCN applied similarly stringent criteria for data homogeneity (like the USHCN) in order to reduce bias. In acquiring data sets, the original observations were sought, and in many cases where bias was identified, the stations in question were removed from the data set. See Section 7 for documentation.

For data collected over the ocean, continuous improvement and greater spatial resolution can be expected in the coming years, with corresponding updates to the historical data. For example, there is a known bias during the World War II years (1941–1945), when almost all ocean temperature measurements were collected by U.S. Navy ships that recorded ocean intake temperatures, which can give warmer results than the techniques used in other years. Future efforts will aim to adjust the data more fully to account for this bias.

nClimDiv Surface Data

The *n*ClimDiv data set follows the same methods as USHCN with regard to detecting and correcting any station biases brought on by changes to the station network over time. Although this data set contains more stations than USHCN, all of the additional stations must meet the same quality criteria as USHCN stations.

Satellite Data

NOAA's satellites cover the entire Earth with consistent measurement methods. Procedures to calibrate the results and correct for any biases over time are described in the references cited under Section 7.

9. Data Limitations

Factors that may impact the confidence, application, or conclusions drawn from this indicator are as follows:

1. Biases in surface measurements may have occurred as a result of changes over time in instrumentation, measuring procedures (e.g., time of day), and the exposure and location of the instruments. Where possible, data have been adjusted to account for changes in these variables. For more information on these corrections, see Section 8. Some scientists believe that the empirical debiasing models used to adjust the data might themselves introduce non-climatic biases (e.g., Pielke et al., 2007).

2. Uncertainties in surface temperature data increase as one goes back in time, as there are fewer stations early in the record. However, these uncertainties are not sufficient to mislead the user about fundamental trends in the data.

10. Sources of Uncertainty

Surface Data

Uncertainties in temperature data increase as one goes back in time, as there are fewer stations early in the record. However, these uncertainties are not sufficient to undermine the fundamental trends in the data.

Error estimates are not readily available for U.S. temperature, but they are available for the global temperature time series. See the error bars in NOAA's graphic online at: www.ncdc.noaa.gov/sotc/service/global/global-land-ocean-mntp-anom/201001-201012.gif. In general, Vose and Menne (2004) suggest that the station density in the U.S. climate network is sufficient to produce a robust spatial average.

Satellite Data

Methods of inferring tropospheric temperature from satellite data have been developed and refined over time. Several independent analyses have produced largely similar curves, suggesting fairly strong agreement and confidence in the results.

Error estimates for the UAH analysis have previously been published in Christy et al. (2000, 2003). Error estimates for the RSS analysis have previously been published in Schabel et al. (2002) and Mears et al. (2003). However, error estimates are not readily available for the updated version of each analysis that EPA obtained in 2014.

11. Sources of Variability

Annual temperature anomalies naturally vary from location to location and from year to year as a result of normal variations in weather patterns, multi-year climate cycles such as the El Niño–Southern Oscillation and Pacific Decadal Oscillation, and other factors. This indicator accounts for these factors by presenting a long-term record (more than a century of data) and averaging consistently over time and space.

12. Statistical/Trend Analysis

This indicator uses ordinary least-squares regression to calculate the slope of the observed trends in temperature. A simple t-test indicates that the following observed trends are significant at the 95 percent confidence level:

- U.S. temperature, 1901-2013, surface: +0.014 °F/year ($p < 0.001$)
- U.S. temperature, 1979-2013, surface: +0.048 °F/year ($p = 0.001$)
- U.S. temperature, 1979-2013, UAH satellite method: +0.041 °F/year ($p < 0.001$)
- U.S. temperature, 1979-2013, RSS satellite method: +0.031 °F/year ($p = 0.006$)
- Global temperature, 1901-2013, surface: +0.015 °F/year ($p < 0.001$)

- Global temperature, 1979-2013, surface: +0.027 °F/year (p < 0.001)
- Global temperature, 1979-2013, UAH satellite method: +0.025 °F/year (p < 0.001)
- Global temperature, 1979-2013, RSS satellite method: +0.023 °F/year (p < 0.001)

To conduct a more complete analysis, however, would potentially require consideration of serial correlation and other more complex statistical factors. Grid cell trends in Figure 3 have not been tested for statistical significance. However, long-term temperature trends in NOAA's 11 climate regions—which are based on state boundaries, with data taken from all complete grid cells and parts of grid cells that fall within these boundaries—*have* been tested for significance. The boundaries of these climate regions and the results of the statistical analysis are reported in the "U.S. and Global Temperature and Precipitation" indicator in EPA's *Report on the Environment*, available at: www.epa.gov/roe. This significance testing provides the basis for the statement in the indicator text that "not all of these regional trends are statistically significant."

References

Balling, Jr., R.C., and C.D. Idso. 2002. Analysis of adjustments to the United States Historical Climatology Network (USHCN) temperature database. Geophys. Res. Lett. 29(10):1387.

Christy, J.R., R.W. Spencer, and W.D. Braswell. 2000. MSU tropospheric temperatures: Dataset construction and radiosonde comparisons. J. Atmos. Ocean. Tech. 17:1153–1170. www.ncdc.noaa.gov/oa/climate/research/uah-msu.pdf.

Christy, J.R., R.W. Spencer, W.B. Norris, W.D. Braswell, and D.E. Parker. 2003. Error estimates of version 5.0 of MSU/AMSU bulk atmospheric temperatures. J. Atmos. Ocean. Tech. 20:613–629.

Karl, T.R., S.J. Hassol, C.D. Miller, and W.L. Murray (eds.). 2006. Temperature trends in the lower atmosphere: Steps for understanding and reconciling differences. U.S. Climate Change Science Program and the Subcommittee on Global Change Research. www.globalchange.gov/browse/reports/sap-11-temperature-trends-lower-atmosphere-steps-understanding-reconciling.

Mears, C.A., M.C. Schabel, and F.J. Wentz. 2003. A reanalysis of the MSU channel 2 tropospheric temperature record. J. Climate 16:3650–3664. www.ncdc.noaa.gov/oa/climate/research/rss-msu.pdf.

Menne, M.J., and C.N. Williams, Jr. 2009. Homogenization of temperature series via pairwise comparisons. J. Climate 22(7):1700–1717.

Menne, M.J., C.N. Williams, Jr., and R.S. Vose. 2009. The U.S. Historical Climatology Network monthly temperature data, version 2. B. Am. Meteorol. Soc. 90:993-1107. ftp://ftp.ncdc.noaa.gov/pub/data/ushcn/v2/monthly/menne-etal2009.pdf.

Peterson, T.C. 2006. Examination of potential biases in air temperature caused by poor station locations. B. Am. Meteorol. Soc. 87:1073–1080. http://journals.ametsoc.org/doi/pdf/10.1175/BAMS-87-8-1073.

Peterson, T.C., R. Vose, R. Schmoyer, and V. Razuvaev. 1998. Global Historical Climatology Network (GHCN) quality control of monthly temperature data. Int. J. Climatol. 18(11):1169–1179.

Pielke, R., J. Nielsen-Gammon, C. Davey, J. Angel, O. Bliss, N. Doesken, M. Cai, S. Fall, D. Niyogi, K. Gallo, R. Hale, K.G. Hubbard, X. Lin, H. Li, and S. Raman. 2007. Documentation of uncertainties and biases associated with surface temperature measurement sites for climate change assessment. B. Am. Meteorol. Soc. 88:913–928.

Schabel, M.C., C.A. Mears, and F.J. Wentz. 2002. Stable long-term retrieval of tropospheric temperature time series from the Microwave Sounding Unit. Proceedings of the International Geophysics and Remote Sensing Symposium III:1845–1847.

Smith, T.M., R.W. Reynolds, T.C. Peterson, and J. Lawrimore. 2008. Improvements to NOAA's historical merged land–ocean surface temperature analysis (1880–2006). J. Climate 21:2283–2296. www.ncdc.noaa.gov/ersst/papers/SEA.temps08.pdf.

Vose, R.S., and M.J. Menne. 2004. A method to determine station density requirements for climate observing networks. J. Climate 17(15):2961–2971.

Vose, R.S., C.N. Williams, Jr., T.C. Peterson, T.R. Karl, and D.R. Easterling. 2003. An evaluation of the time of observation bias adjustment in the U.S. Historical Climatology Network. Geophys. Res. Lett. 30(20):2046.

Watts, A. 2009. Is the U.S. surface temperature record reliable? The Heartland Institute. http://wattsupwiththat.files.wordpress.com/2009/05/surfacestationsreport_spring09.pdf.

Willmott, C.J., and S.M. Robeson. 1995. Climatologically aided interpolation (CAI) of terrestrial air temperature. Int. J. Climatol. 15(2):221–229.

High and Low Temperatures

Identification

1. Indicator Description

This indicator examines metrics related to trends in unusually hot and cold temperatures across the United States over the last several decades. Changes in many extreme weather and climate events have been observed, and further changes, such as fewer cold days and nights and more frequent hot days and nights, are likely to continue this century (IPCC, 2013). Extreme temperature events like summer heat waves and winter cold spells can have profound effects on human health and society.

Components of this indicator include:

- An index reflecting the frequency of extreme heat events (Figure 1).

- The percentage of land area experiencing unusually hot summer temperatures or unusually cold winter temperatures (Figures 2 and 3, respectively).

- Changes in the prevalence of unusually hot and unusually cold temperatures throughout the year at individual weather stations (Figures 4 and 5).

- The proportion of record-setting high temperatures to record low temperatures over time (Figure 6).

2. Revision History

April 2010:   Indicator posted.
December 2011:   Updated Figure 1 with data through 2010; combined Figures 2 and 3 into a new
                        Figure 2, and updated data through 2011; added new Figures 3 and 4 ("Unusually
                        Cold Winter Temperatures" and "Record Daily Highs and Record Daily Lows"); and
                        expanded the indicator from "Heat Waves" to "High and Low Temperatures."
February 2012:   Updated Figure 1 with data through 2011.
March 2012:   Updated Figure 3 with data through 2012.
October 2012:   Updated Figure 2 with data through 2012.
August 2013:   Updated Figure 1 on EPA's website with data through 2012; updated Figure 3 on
                        EPA's website with data through 2013.
December 2013:   Added new percentile metrics as Figures 4 and 5; changed "Record Daily Highs and
                        Record Daily Lows" to Figure 6.
April 2014:   Updated Figures 1, 2, 4, and 5 with data through 2013; updated Figure 3 with data
                        through 2014.

3. Data Sources

Index values for Figure 1 were provided by Dr. Kenneth Kunkel of the National Oceanic and Atmospheric Administration's (NOAA's) Cooperative Institute for Climate and Satellites (CICS), who updated an analysis that was previously published in U.S. Climate Change Science Program (2008). Data for Figures 2 and 3 come from NOAA's U.S. Climate Extremes Index (CEI), which is maintained by NOAA's National Climatic Data Center (NCDC) and based on the U.S. Historical Climatology Network (USHCN). Data for Figures 4 and 5 come from U.S. weather stations within NCDC's Global Historical Climatology Network (GHCN). Data for Figure 6 come from an analysis published by Meehl et al. (2009).

All components of this indicator are based on temperature measurements from weather stations overseen by NOAA's National Weather Service (NWS). These underlying data are maintained by NCDC.

4. Data Availability

Figure 1. U.S. Annual Heat Wave Index, 1895–2013

Data for this figure were provided by Dr. Kenneth Kunkel of NOAA CICS, who performed the analysis based on data from NCDC's publicly available databases.

Figures 2 and 3. Area of the Contiguous 48 States with Unusually Hot Summer Temperatures (1910–2013) or Unusually Cold Winter Temperatures (1911–2014)

NOAA has calculated each of the components of the CEI and has made these data files publicly available. The data for unusually hot summer maximum and minimum temperatures (CEI steps 1b and 2b) can be downloaded from: ftp://ftp.ncdc.noaa.gov/pub/data/cei/dk-step1-hi.06-08.results and: ftp://ftp.ncdc.noaa.gov/pub/data/cei/dk-step2-hi.06-08.results, respectively. The data for unusually cold winter maximum and minimum temperatures (CEI steps 1a and 2a) can be downloaded from: ftp://ftp.ncdc.noaa.gov/pub/data/cei/dk-step1-lo.12-02.results and: ftp://ftp.ncdc.noaa.gov/pub/data/cei/dk-step2-lo.12-02.results, respectively. A "readme" file (ftp://ftp.ncdc.noaa.gov/pub/data/cei) explains the contents of the data files. NOAA's CEI website (www.ncdc.noaa.gov/extremes/cei) provides additional descriptions and links, along with a portal to download or graph various components of the CEI, including the data sets listed above.

Figures 4 and 5. Changes in Unusually Hot and Cold Temperatures in the Contiguous 48 States, 1948–2013

Data for these maps came from Version 3.12 of NCDC's GHCN-Daily data set, which provided the optimal format for processing. Within the contiguous 48 states, the GHCN pulls data directly from a dozen separate data sets maintained at NCDC, including the USHCN. NCDC explains the variety of databases that feed into the GHCN for U.S.-based stations in online metadata and at: www.ncdc.noaa.gov/oa/climate/ghcn-daily/index.php?name=source. The data for this indicator can be obtained online via FTP at: ftp://ftp.ncdc.noaa.gov/pub/data/ghcn/daily/hcn. Appropriate metadata and "readme" files are available at: ftp://ftp.ncdc.noaa.gov/pub/data/ghcn/daily.

Figure 6. Record Daily High and Low Temperatures in the Contiguous 48 States, 1950–2009

Ratios of record highs to lows were taken from Meehl et al. (2009) and a supplemental release that accompanied the publication of that peer-reviewed study (www2.ucar.edu/atmosnews/news/1036/record-high-temperatures-far-outpace-record-lows-across-us). Meehl et al. (2009) covered the period 1950–2006, so the "2000s" bar in Figure 6 is based on a subsequent analysis of data through 2009 that was conducted by the authors of the paper and presented in the aforementioned press release. For confirmation, EPA obtained the actual counts of highs and lows by decade from Claudia Tebaldi, a co-author of the Meehl et al. (2009) paper.

Underlying Data

NCDC maintains a set of databases that provide public access to daily and monthly temperature records from thousands of weather stations across the country. For access to these data and accompanying metadata, see NCDC's website at: www.ncdc.noaa.gov.

Many of the weather stations are part of NOAA's Cooperative Observer Program (COOP). Complete data, embedded definitions, and data descriptions for these stations can be found online at: www.ncdc.noaa.gov/doclib. State-specific data can be found at: www7.ncdc.noaa.gov/IPS/coop/coop.html;jsessionid=312EC0892FFC2FBB78F63D0E3ACF6CBC. There are no confidentiality issues that may limit accessibility. Additional metadata can be found at: www.nws.noaa.gov/om/coop.

Methodology

5. Data Collection

Since systematic collection of weather data in the United States began in the 1800s, observations have been recorded from 23,000 stations. At any given time, approximately 8,000 stations are recording observations on an hourly basis, along with the maximum and minimum temperatures for each day.

NOAA's National Weather Service (NWS) operates some stations (called first-order stations), but the large majority of U.S. weather stations are part of NWS's Cooperative Observer Program (COOP). The COOP data set is the core climate network of the United States (Kunkel et al., 2005). Cooperative observers include state universities, state and federal agencies, and private individuals. Observers are trained to collect data following NWS protocols, and equipment to gather these data is provided and maintained by the NWS.

Data collected by COOP are referred to as U.S. Daily Surface Data or Summary of the Day data. Variables that are relevant to this indicator include observations of daily maximum and minimum temperatures. General information about the NWS COOP data set is available at: www.nws.noaa.gov/os/coop/what-is-coop.html. Sampling procedures are described in Kunkel et al. (2005) and in the full metadata for the COOP data set, available at: www.nws.noaa.gov/om/coop.

NCDC also maintains the USHCN, on which the CEI is based. The USHCN contains data from a subset of COOP and first-order weather stations that meet certain selection criteria and undergo additional levels of quality control. The USHCN contains daily and monthly averaged maximum, minimum, and mean temperature data from approximately 1,200 stations within the contiguous 48 states. The period of record varies for each station, but generally includes most of the 20th century. One of the objectives in establishing the USHCN was to detect secular changes in regional rather than local climate. Therefore, stations included in this network are only those believed to not be influenced to any substantial degree by artificial changes of local environments. To be included in the USHCN, a station had to meet certain criteria for record longevity, data availability (percentage of available values), spatial coverage, and consistency of location (i.e., experiencing few station changes). An additional criterion, which sometimes compromised the preceding criteria, was the desire to have a uniform distribution of stations across the United States. Included with the data set are metadata files that contain information about station moves, instrumentation, observing times, and elevation. NOAA's website (www.ncdc.noaa.gov/oa/climate/research/ushcn) provides more information about USHCN data collection.

GHCN-Daily Version 3.12 contains historical daily weather data from 49,422 monitoring stations across the United States. Temperature data were obtained from many types of stations. NCDC has published documentation for the GHCN. For more information, including data sources, methods, and recent updates to the data set, see the metadata and "readme" files at: ftp://ftp.ncdc.noaa.gov/pub/data/ghcn/daily.

All six figures use data from the contiguous 48 states. Original sources and selection criteria are as follows:

- Figure 1 is based on stations from the COOP data set that had sufficient data during the period of record (1895–2013).

- Figures 2 and 3 are based on the narrower set of stations contained within the USHCN, which is the source of all data for NOAA's CEI. Additional selection criteria were applied to these data prior to inclusion in CEI calculations, as described by Gleason et al. (2008). In compiling the temperature components of the CEI, NOAA selected only those stations with monthly temperature data at least 90 percent complete within a given period (e.g., annual, seasonal) as well as 90 percent complete for the full period of record.

- Figures 4 and 5 use daily maximum and minimum temperature data from the GHCN-Daily. This analysis is limited to the period from 1948 to 2013 because it enabled inclusion of most stations from the USHCN, which is a key contributing database to the GHCN-Daily. Station data are included only for years in which data are reported (one or more days) in six or more months. If a station reported data from fewer than six months, data from the entire year are removed. After filtering for individual years (above), stations are removed from further consideration if fewer than 48 years of data are included. Years need not be consecutive. As a result, Figures 4 and 5 show trends for 1,119 stations.

- In Figure 6, data for the 1950s through 1990s are based on a subset of 2,000 COOP stations that have collected data since 1950 and had no more than 10 percent missing values during the period from 1950 to 2006. These selection criteria are further described in Meehl et al. (2009).

- In Figure 6, data for the 2000s are based on the complete set of COOP records available from 2000 through September 2009. These numbers were published in Meehl et al. (2009) and the accompanying press release, but they do not follow the same selection criteria as the previous decades (as described above). Counts of record highs and lows using the Meehl et al. (2009) selection criteria were available, but only through 2006. Thus, to make this indicator as current as possible, EPA chose to use data from the broader set that extends through September 2009. Using the 2000–2006 data would result in a high:low ratio of 1.86, compared with a ratio of 2.04 when the full-decade data set (shown in Figure 6) is considered.

6. Indicator Derivation

Figure 1. U.S. Annual Heat Wave Index, 1895–2013

Data from the COOP data set have been used to calculate annual values for a U.S. Annual Heat Wave Index. In this indicator, heat waves are defined as warm periods of at least four days with an average temperature (that is, averaged over all four days) exceeding the threshold for a one-in-10-year occurrence (Kunkel et al., 1999). The Annual U.S. Heat Wave Index is a frequency measure of the number of heat waves that occur each year. A complete explanation of trend analysis in the annual average heat wave index values, especially trends occurring since 1960, can be found in Appendix A, Example 2, of U.S. Climate Change Science Program (2008). Analytical procedures are described in Kunkel et al. (1999).

Figures 2 and 3. Area of the Contiguous 48 States with Unusually Hot Summer Temperatures (1910–2013) or Unusually Cold Winter Temperatures (1911–2014)

Figure 2 of this indicator shows the percentage of the area of the contiguous 48 states in any given year that experienced unusually warm maximum and minimum summer temperatures. Figure 3 displays the percentage of land area that experienced unusually cold maximum and minimum winter temperatures.

Figures 2 and 3 were developed as subsets of NOAA's CEI, an index that uses six variables to examine trends in extreme weather and climate. These figures are based on components of NOAA's CEI (labeled as Steps 1a, 1b, 2a, and 2b) that look at the percentage of land area within the contiguous 48 states that experienced maximum (Step 1) or minimum (Step 2) temperatures much below (a) or above (b) normal.

NOAA computed the data for the CEI and calculated the percentage of land area for each year by dividing the contiguous 48 states into a 1-degree by 1-degree grid and using data from one station per grid box. This was done to eliminate many of the artificial extremes that resulted from a changing number of available stations over time.

NOAA began by averaging all daily highs at a given station over the course of a month to derive a monthly average high, then performing the same step with daily lows. Next, period (monthly) averages were sorted and ranked, and values were identified as "unusually warm" if they fell in the highest 10th percentile in the period of record for each station or grid cell, and "unusually cold" if they fell in the lowest 10th percentile. Thus, the CEI has been constructed to have an expected value of 10 percent for each of these components, based on the historical record—or a value of 20 percent if the two extreme ends of the distribution are added together.

The CEI can be calculated for individual months, seasons, or an entire year. Figure 2 displays data for summer, which the CEI defines as June, July, and August. Figure 3 displays data for winter, which the CEI defines as December, January, and February. Winter values are plotted at the year in which the season ended; for example, the winter from December 2013 to February 2014 is plotted at year 2014. This explains why Figures 2 and 3 appear to have a different starting year, as data were not available from December 1909 to calculate a winter value for 1910. To smooth out some of the year-to-year variability, EPA applied a nine-point binomial filter, which is plotted at the center of each nine-year window. For example, the smoothed value from 2006 to 2014 is plotted at year 2010. NOAA NCDC recommends this approach and has used it in the official online reporting tool for the CEI.

EPA used endpoint padding to extend the nine-year smoothed lines all the way to the ends of the period of record. As recommended by NCDC, EPA calculated smoothed values as follows: if 2014 was the most recent year with data available, EPA calculated smoothed values to be centered at 2011, 2012, 2013, and 2014 by inserting the 2014 data point into the equation in place of the as-yet-unreported annual data points for 2015 and beyond. EPA used an equivalent approach at the beginning of the time series.

The CEI has been extensively documented and refined over time to provide the best possible representation of trends in extreme weather and climate. For an overview of how NOAA constructed Steps 1 and 2 of the CEI, see: www.ncdc.noaa.gov/extremes/cei. This page provides a list of references that describe analytical methods in greater detail. In particular, see Gleason et al. (2008).

Figures 4 and 5. Changes in Unusually Hot and Cold Temperatures in the Contiguous 48 States, 1948– 2013

For Figure 4, the change in the number of days per year on which the daily maximum temperature exceeded the 95th percentile temperature was determined through the following steps:

1. At each monitoring station, the 95th percentile daily maximum temperature was determined for the full period of record (1948–2013).

2. For each station, the number of days in each calendar year on which the maximum daily temperature exceeded the station-specific 95th percentile temperature was determined.

3. The average rate of change over time in the number of >95th percentile days was estimated from the annual number of >95th percentile days using ordinary least-squares linear regression.

4. Regression coefficients (the average change per year in >95th percentile days) for regressions significant at the 90 percent level ($p \leq 0.1$) were multiplied by the number of years in the analysis (1948–2013 = 65 years) to estimate the total change in the number of annual >95th percentile days over the full period of record. Where $p > 0.1$, coefficients were set to zero. These values (including "zero" values for stations with insignificant trends) were mapped to show trends at each climate station.

Figure 5 was constructed using a similar procedure with daily minimum temperatures and the 5th percentile.

Figure 6. Record Daily High and Low Temperatures in the Contiguous 48 States, 1950–2009

Figure 6 displays the proportion of daily record high and daily record low temperatures reported at a subset of quality-controlled NCDC COOP network stations (except for the most recent decade, which is based on the entire COOP network, as described in Section 5). As described in Meehl et al. (2009), steps were taken to fill missing data points with simple averages from neighboring days with reported values when there are no more than two consecutive days missing, or otherwise by interpolating values at the closest surrounding stations.

Based on the total number of record highs and the total number of record lows set in each decade, Meehl et al. (2009) calculated each decade's ratio of record highs to record lows. EPA converted these values to percentages to make the results easier to communicate.

Although it might be interesting to look at trends in the absolute number of record highs and record lows over time, these values are recorded in a way that would make a trend analysis misleading. A daily high or low is registered as a "record" if it broke a record *at the time*—even if that record has since been surpassed. Statistics dictate that as more years go by, it becomes less likely that a record will be broken. In contrast, if a station has only been measuring temperature for 5 years (for example), every day has a much greater chance of breaking a previous record. Thus, a decreasing trend in absolute counts does not indicate that the climate is actually becoming less extreme, as one might initially guess. Meehl et al. (2009) show that actual counts indeed fit a decreasing pattern over time, as expected statistically.

7. Quality Assurance and Quality Control

The NWS has documented COOP methods, including training manuals and maintenance of equipment, at: www.nws.noaa.gov/os/coop/training.htm. These training materials also discuss quality control of the underlying data set. Additionally, pre-1948 data in the COOP data set have recently been digitized from hard copies. Quality control procedures associated with digitization and other potential sources of error are discussed in Kunkel et al. (2005).

Quality control procedures for the USHCN are summarized at: www.ncdc.noaa.gov/oa/climate/research/ushcn/#processing. Homogeneity testing and data correction methods are described in numerous peer-reviewed scientific papers by NCDC. A series of data corrections was developed to specifically address potential problems in trend estimation of the rates of warming or cooling in USHCN Version 2. They include:

- Removal of duplicate records.
- Procedures to deal with missing data.
- Adjusting for changes in observing practices, such as changes in observation time.
- Testing and correcting for artificial discontinuities in a local station record, which might reflect station relocation, instrumentation changes, or urbanization (e.g., heat island effects).

Quality control procedures for GHCN-Daily data are described at: www.ncdc.noaa.gov/oa/climate/ghcn-daily/index.php?name=quality. GHCN-Daily data undergo rigorous quality assurance reviews, starting with pre-screening for data and station appropriateness. This data integration process is described in detail at: www.ncdc.noaa.gov/oa/climate/ghcn-daily/index.php?name=integration. Further quality assurance procedures for individual data points include removal of duplicates, isolated values, suspicious streaks, and excessive or unnatural values; spatial comparisons that verify the accuracy of the climatological mean and the seasonal cycle; and neighbor checks that identify outliers from both a serial and a spatial perspective. Data that fail a given quality control check (0.3 percent of all values) are marked with flags, depending on the type of error identified.

Analysis

8. Comparability Over Time and Space

Long-term weather stations have been carefully selected from the full set of all COOP stations to provide an accurate representation of the United States for the U.S. Annual Heat Wave Index and the proportion of record daily highs to record daily lows (Kunkel et al., 1999; Meehl et al., 2009). Some bias may have occurred as a result of changes over time in instrumentation, measuring procedures, and the exposure and location of the instruments. The record high/low analysis begins with 1950 data, in an effort to reduce disparity in station record lengths.

The USHCN has undergone extensive testing to identify errors and biases in the data and either remove these stations from the time series or apply scientifically appropriate correction factors to improve the utility of the data. In particular, these corrections address changes in the time-of-day of observation, advances in instrumentation, and station location changes.

Homogeneity testing and data correction methods are described in more than a dozen peer-reviewed scientific papers by NCDC. Data corrections were developed to specifically address potential problems in trend estimation of the rates of warming or cooling in the USHCN (see Section 7 for documentation). Balling and Idso (2002) compared the USHCN data with several surface and upper-air data sets and showed that the effects of the various USHCN adjustments produce a significantly more positive, and likely spurious, trend in the USHCN data. However, Balling and Idso (2002) drew conclusions based on an analysis that is no longer valid, as it relied on the UAH satellite temperature data set before corrections identified by Karl et al. (2006) were applied to the satellite record. These corrections have been accepted by all the researchers involved, including those at UAH, and they increased the temperature trend in the satellite data set, eliminating many of the discrepancies with the surface temperature data set. Additionally, even before these corrections were identified, Vose et al. (2003) found that "the time of observation bias adjustments in HCN appear to be robust," contrary to the assertions of Balling and Idso that these adjustments were biased. Vose et al. (2003) found that USHCN station history information is reasonably complete and that the bias adjustment models have low residual errors.

Further analysis by Menne et al. (2009) suggests that:

> ...the collective impact of changes in observation practice at USHCN stations is systematic and of the same order of magnitude as the background climate signal. For this reason, bias adjustments are essential to reducing the uncertainty in U.S. climate trends. The largest biases in the HCN are shown to be associated with changes to the time of observation and with the widespread changeover from liquid-in-glass thermometers to the maximum minimum temperature sensor (MMTS). With respect to [USHCN] Version 1, Version 2 trends in maximum temperatures are similar while minimum temperature trends are somewhat smaller because of an apparent overcorrection in Version 1 for the MMTS instrument change, and because of the systematic impact of undocumented station changes, which were not addressed [in] Version 1.

USHCN Version 2 represents an improvement in this regard.

Some observers have expressed concerns about other aspects of station location and technology. For example, Watts (2009) expresses concern that many U.S. weather stations are sited near artificial heat sources such as buildings and paved areas, potentially biasing temperature trends over time. In response to these concerns, NOAA analyzed trends for a subset of stations that Watts had determined to be "good or best," and found the temperature trend over time to be very similar to the trend across the full set of USHCN stations (www.ncdc.noaa.gov/oa/about/response-v2.pdf). NOAA's Climate Reference Network (www.ncdc.noaa.gov/crn), a set of optimally-sited stations completed in 2008, can be used to test the accuracy of recent trends. While it is true that many other stations are not optimally located, NOAA's findings support the results of an earlier analysis by Peterson (2006), who found no significant bias in long-term trends associated with station siting once NOAA's homogeneity adjustments were applied. An independent analysis by the Berkeley Earth Surface Temperature (BEST) project (http://berkeleyearth.org/summary-of-findings) used more stations and a different statistical methodology, yet found similar results.

As documented in Section 7, GHCN-Daily stations are extensively filtered and quality-controlled to maximize comparability across stations and across time.

9. Data Limitations

Factors that may impact the confidence, application, or conclusions drawn from this indicator are as follows:

1. Biases may have occurred as a result of changes over time in instrumentation, measuring procedures, and the exposure and location of the instruments. Where possible, data have been adjusted to account for changes in these variables. For more information on these corrections, see Section 7. Some scientists believe that the empirical debiasing models used to adjust the data might themselves introduce non-climatic biases (e.g., Pielke et al., 2007).

2. Observer errors, such as errors in reading instruments or writing observations on the form, are present in the earlier part of this data set. Additionally, uncertainty may be introduced into this data set when hard copies of data are digitized. As a result of these and other factors, uncertainties in the temperature data increase as one goes back in time, particularly because there were fewer stations early in the record. However, NOAA does not believe these uncertainties are sufficient to undermine the fundamental trends in the data. More information about limitations of pre-1948 weather data can be found in Kunkel et al. (2005).

10. Sources of Uncertainty

Uncertainty may be introduced into this data set when hard copies of historical data are digitized. For this and other reasons, uncertainties in the temperature data increase as one goes back in time, particularly because there are fewer stations early in the record. However, NOAA does not believe these uncertainties are sufficient to undermine the fundamental trends in the data. Vose and Menne (2004) suggest that the station density in the U.S. climate network is sufficient to produce robust spatial averages.

Error estimates have been developed for certain segments of the data set, but do not appear to be available for the data set as a whole. Uncertainty measurements are not included with the publication of the U.S. Annual Heat Wave Index or the CEI seasonal temperature data. Error measurements for the pre-1948 COOP data set are discussed in detail in Kunkel et al. (2005).

11. Sources of Variability

Inter-annual temperature variability results from normal year-to-year variation in weather patterns, multi-year climate cycles such as the El Niño–Southern Oscillation and Pacific Decadal Oscillation, and other factors. This indicator presents nine-year smoothed curves (Figures 1, 2, and 3), long-term rates of change (Figures 4 and 5), and decadal averages (Figure 6) to reduce the year-to-year "noise" inherent in the data. Temperature patterns also vary spatially. This indicator provides information on geographic differences using location-specific trends in Figures 4 and 5.

12. Statistical/Trend Analysis

Heat wave trends (Figure 1) are somewhat difficult to analyze because of several outlying values in data from the 1930s. Statistical methods used to analyze trends in the U.S. Annual Heat Wave Index are presented in Appendix A, Example 2, of U.S. Climate Change Science Program (2008). Despite the presence of inter-annual variability and several outlying values in the 1930s, standard statistical treatments can be applied to assess a highly statistically significant linear trend from 1960 to 2011. However, the trend over the full period of record is not statistically significant.

Figures 4 and 5 use ordinary least-squares linear regression to calculate the slope of observed trends in the annual number of 95th and 5th percentile days at each monitoring station. Trends that are not statistically significant at the 90 percent level ($p \leq 0.1$) are displayed as zero (i.e., they are grouped into the "-5 to 5" class).

This indicator does not report on the slope of the apparent trends in Figures 2, 3, and 6, nor does it calculate the statistical significance of these trends.

References

Balling, Jr., R.C., and C.D. Idso. 2002. Analysis of adjustments to the United States Historical Climatology Network (USHCN) temperature database. Geophys. Res. Lett. 29(10):1387.

Gleason, K.L., J.H. Lawrimore, D.H. Levinson, T.R. Karl, and D.J. Karoly. 2008. A revised U.S. climate extremes index. J. Climate 21:2124–2137.

IPCC (Intergovernmental Panel on Climate Change). 2013. Climate change 2013: The physical science basis. Summary for policymakers. Working Group I contribution to the IPCC Fifth Assessment Report. Cambridge, United Kingdom: Cambridge University Press. www.ipcc.ch/report/ar5/wg1.

Kunkel, K.E., R.A. Pielke Jr., and S. A. Changnon. 1999. Temporal fluctuations in weather and climate extremes that cause economic and human health impacts: A review. B. Am. Meteorol. Soc. 80:1077–1098.

Kunkel, K.E., D.R. Easterling, K. Hubbard, K. Redmond, K. Andsager, M.C. Kruk, and M.L. Spinar. 2005. Quality control of pre-1948 Cooperative Observer Network data. J. Atmos. Ocean. Tech. 22:1691–1705.

Meehl, G.A., C. Tebaldi, G. Walton, D. Easterling, and L. McDaniel. 2009. Relative increase of record high maximum temperatures compared to record low minimum temperatures in the U.S. Geophys. Res. Lett. 36:L23701.

Menne, M.J., C.N. Williams, Jr., and R.S. Vose. 2009. The U.S. Historical Climatology Network monthly temperature data, version 2. B. Am. Meteorol. Soc. 90:993–1107.
ftp://ftp.ncdc.noaa.gov/pub/data/ushcn/v2/monthly/menne-etal2009.pdf.

Peterson, T.C. 2006. Examination of potential biases in air temperature caused by poor station locations. B. Am. Meteorol. Soc. 87:1073–1080. http://journals.ametsoc.org/doi/pdf/10.1175/BAMS-87-8-1073.

Pielke, R., J. Nielsen-Gammon, C. Davey, J. Angel, O. Bliss, N. Doesken, M. Cai, S. Fall, D. Niyogi, K. Gallo, R. Hale, K.G. Hubbard, X. Lin, H. Li, and S. Raman. 2007. Documentation of uncertainties and biases associated with surface temperature measurement sites for climate change assessment. B. Am. Meteorol. Soc. 88:913–928.

U.S. Climate Change Science Program. 2008. Synthesis and Assessment Product 3.3: Weather and climate extremes in a changing climate. http://library.globalchange.gov/products/assessments/2004-2009-synthesis-and-assessment-products/sap-3-3-weather-and-climate-extremes-in-a-changing-climate.

Vose, R.S., and M.J. Menne. 2004. A method to determine station density requirements for climate observing networks. J. Climate 17(15):2961–2971.

Vose, R.S., C.N. Williams, Jr., T.C. Peterson, T.R. Karl, and D.R. Easterling. 2003. An evaluation of the time of observation bias adjustment in the U.S. Historical Climatology Network. Geophys. Res. Lett. 30(20):2046.

Watts, A. 2009. Is the U.S. surface temperature record reliable? The Heartland Institute. http://wattsupwiththat.files.wordpress.com/2009/05/surfacestationsreport_spring09.pdf.

U.S. and Global Precipitation

Identification

1. Indicator Description

This indicator describes changes in total precipitation over land for the United States and the world from 1901 to 2012. In this indicator, precipitation data are presented as trends in anomalies. Precipitation is an important component of climate, and changes in precipitation can have wide-ranging direct and indirect effects on the environment and society. As average temperatures at the Earth's surface rise, more evaporation occurs, which, in turn, increases overall precipitation. Therefore, a warming climate is expected to increase precipitation in many areas. However, factors such as shifting wind patterns and changes in the ocean currents that drive the world's climate system will also cause some areas to experience decreased precipitation.

Components of this indicator include:

- Changes in precipitation in the contiguous 48 states over time (Figure 1).
- Changes in worldwide precipitation over land through time (Figure 2).
- A map showing rates of precipitation change across the United States (Figure 3).

2. Revision History

April 2010:   Indicator posted.
December 2011:   Updated with data through 2010.
May 2012:   Updated with data through 2011.
August 2013:   Updated indicator on EPA's website with data through 2012.

Data Sources

3. Data Sources

This indicator is based on precipitation anomaly data provided by the National Oceanic and Atmospheric Administration's (NOAA's) National Climatic Data Center (NCDC).

4. Data Availability

Data for this indicator were provided to EPA by NOAA's NCDC. NCDC calculated these time series based on monthly values from two NCDC-maintained databases: the U.S. Historical Climatology Network (USHCN) Version 2.5 and the Global Historical Climatology Network–Monthly (GHCN-M) Version 2. Both of these databases can be accessed online.

Contiguous 48 States

Underlying precipitation data for the contiguous 48 states come from the USHCN. Currently, the data are distributed by NCDC on various computer media (e.g., anonymous FTP sites), with no confidentiality issues limiting accessibility. Users can link to the data online at: www.ncdc.noaa.gov/oa/climate/research/ushcn. Appropriate metadata and "readme" files are appended to the data. For example, see: ftp://ftp.ncdc.noaa.gov/pub/data/ushcn/v2/monthly/readme.txt.

Alaska, Hawaii, and Global

GHCN precipitation data can be obtained from NCDC over the Web or via anonymous FTP. For access to GHCN data, see: www.ncdc.noaa.gov/ghcnm/v2.php. There are no known confidentiality issues that limit access to the data set, and the data are accompanied by metadata.

Methodology

5. Data Collection

This indicator is based on precipitation measurements collected from thousands of land-based weather stations throughout the United States and worldwide, using standard meteorological instruments. Data for the contiguous 48 states were compiled in the USHCN. Data for Alaska, Hawaii, and the rest of the world were taken from the GHCN. Both of these networks are overseen by NOAA and have been extensively peer reviewed. As such, they represent the most complete long-term instrumental data sets for analyzing recent climate trends. More information on these networks can be found below.

Contiguous 48 States

USHCN Version 2.5 contains total monthly precipitation data from approximately 1,200 stations within the contiguous 48 states. The period of record varies for each station, but generally includes most of the 20th century. One of the objectives in establishing the USHCN was to detect secular changes in regional rather than local climate. Therefore, stations included in the network are only those believed to not be influenced to any substantial degree by artificial changes of local environments. Some of the stations in the USHCN are first-order weather stations, but the majority are selected from U.S. cooperative weather stations (approximately 5,000 in the United States). To be included in the USHCN, a station had to meet certain criteria for record longevity, data availability (percentage of available values), spatial coverage, and consistency of location (i.e., experiencing few station changes). An additional criterion, which sometimes compromised the preceding criteria, was the desire to have a uniform distribution of stations across the United States. Included with the data set are metadata files that contain information about station moves, instrumentation, observing times, and elevation. NOAA's website (www.ncdc.noaa.gov/oa/climate/research/ushcn) provides more information about USHCN data collection.

Alaska, Hawaii, and Global

GHCN Version 2 contains monthly climate data from 20,590 weather stations worldwide. Data were obtained from many types of stations.

NCDC has published documentation for the GHCN. For more information, including data sources, methods, and recent improvements, see: www.ncdc.noaa.gov/ghcnm/v2.php and the sources listed therein.

6. Indicator Derivation

NOAA calculated monthly precipitation totals for each site. In populating the USHCN and GHCN, NOAA employed a homogenization algorithm to identify and correct for substantial shifts in local-scale data that might reflect changes in instrumentation, station moves, or urbanization effects. These adjustments were performed according to published, peer-reviewed methods. For more information on these quality assurance and error correction procedures, see Section 7.

In this indicator, precipitation data are presented as trends in anomalies. An anomaly is the difference between an observed value and the corresponding value from a baseline period. This indicator uses a baseline period of 1901 to 2000. The choice of baseline period will not affect the shape or the statistical significance of the overall trend in anomalies. For precipitation (percentage anomalies), choosing a different baseline period would move the curve up or down and possibly change the magnitude slightly.

To generate the precipitation time series, NOAA converted measurements into anomalies for total monthly precipitation in millimeters. Monthly anomalies were added to find an annual anomaly for each year, which was then converted to a percent anomaly—i.e., the percent departure from the average annual precipitation during the baseline period.

To achieve uniform spatial coverage (i.e., not biased toward areas with a higher concentration of measuring stations), NOAA averaged anomalies within grid cells on the map to create "gridded" data sets. The graph for the contiguous 48 states (Figure 1) and the map (Figure 3) are based on an analysis using grid cells that measure 2.5 degrees latitude by 3.5 degrees longitude. The global graph (Figure 2) comes from an analysis of grid cells measuring 5 degrees by 5 degrees. These particular grid sizes have been determined to be optimal for analyzing USHCN and GHCN climate data; see: www.ncdc.noaa.gov/oa/climate/research/ushcn/gridbox.html for more information.

Figures 1 and 2 show trends from 1901 to 2012, based on NOAA's gridded data sets. Although earlier data are available for some stations, 1901 was selected as a consistent starting point.

The map in Figure 3 shows long-term rates of change in precipitation over the United States for the 1901–2012 period, except for Alaska and Hawaii, for which widespread and reliable data collection did not begin until 1918 and 1905, respectively. A regression was performed on the annual anomalies for each grid cell. Trends were calculated only in those grid cells for which data were available for at least 66 percent of the years during the full period of record. The slope of each trend (percent change in precipitation per year) was calculated from the annual time series by ordinary least-squares regression and then multiplied by 100 to obtain a rate per century. No attempt has been made to portray data beyond the time and space in which measurements were made.

NOAA is continually refining historical data points in the USHCN and GHCN, often as a result of improved methods to reduce bias and exclude erroneous measurements. These improvements frequently result in the designation of new versions of the USHCN and GHCN. As EPA updates this indicator to reflect these upgrades, slight changes to some historical data points may become apparent.

7. Quality Assurance and Quality Control

Both the USHCN and the GHCN have undergone extensive quality assurance procedures to identify errors and biases in the data and to remove these stations from the time series or apply correction factors.

Contiguous 48 States

Quality control procedures for the USHCN are summarized at: www.ncdc.noaa.gov/oa/climate/research/ushcn. Homogeneity testing and data correction methods are described in numerous peer-reviewed scientific papers by NOAA's NCDC. A series of data corrections was developed to address specific potential problems in trend estimation in USHCN Version 2.5. They include:

- Removal of duplicate records.
- Procedures to deal with missing data.
- Testing and correcting for artificial discontinuities in a local station record, which might reflect station relocation or instrumentation changes.

Alaska, Hawaii, and Global

QA/QC procedures for GHCN precipitation data are described at: www.ncdc.noaa.gov/ghcnm/v2.php. GHCN data undergo rigorous quality assurance reviews, which include pre-processing checks on source data; removal of duplicates, isolated values, and suspicious streaks; time series checks to identify spurious changes in the mean and variance; spatial comparisons to verify the accuracy of the climatological mean and the seasonal cycle; and neighbor checks to identify outliers from both a serial and a spatial perspective.

Analysis

8. Comparability Over Time and Space

Both the USHCN and the GHCN have undergone extensive testing to identify errors and biases in the data and either remove these stations from the time series or apply scientifically appropriate correction factors to improve the utility of the data. In particular, these corrections address advances in instrumentation and station location changes. See Section 7 for documentation.

9. Data Limitations

Factors that may impact the confidence, application, or conclusions drawn from this indicator are as follows:

1. Biases in measurements may have occurred as a result of changes over time in instrumentation, measuring procedures, and the exposure and location of the instruments. Where possible, data have been adjusted to account for changes in these variables. For more information on these corrections, see Section 7.

2. As noted in Section 10, uncertainties in precipitation data increase as one goes back in time, as there are fewer stations early in the record. However, these uncertainties are not sufficient to undermine the fundamental trends in the data.

10. Sources of Uncertainty

Uncertainties in precipitation data increase as one goes back in time, as there are fewer stations early in the record. However, these uncertainties are not sufficient to undermine the fundamental trends in the data.

Error estimates are not readily available for U.S. or global precipitation. Vose and Menne (2004) suggest that the station density in the U.S. climate network is sufficient to produce a robust spatial average.

11. Sources of Variability

Annual precipitation anomalies naturally vary from location to location and from year to year as a result of normal variation in weather patterns, multi-year climate cycles such as the El Niño–Southern Oscillation and Pacific Decadal Oscillation, and other factors. This indicator accounts for these factors by presenting a long-term record (more than a century of data) and averaging consistently over time and space.

12. Statistical/Trend Analysis

This indicator uses ordinary least-squares regression to calculate the slope of the observed trends in precipitation. A simple t-test indicates that the following observed trends are significant at the 95 percent confidence level:

- U.S. precipitation, 1901-2012: +0.050 %/year ($p = 0.010$).
- Global precipitation, 1901-2012: +0.022 %/year ($p < 0.001$).

To conduct a more complete analysis, however, would potentially require consideration of serial correlation and other more complex statistical factors. Grid cell trends in Figure 3 have not been tested for statistical significance. However, long-term precipitation trends in NOAA's 11 climate regions—which are based on state boundaries, with data taken from all complete grid cells and portions of grid cells that fall within these boundaries—*have* been tested for significance. The boundaries of these climate regions and the results of the statistical analysis are reported in the "U.S. and Global Temperature and Precipitation" indicator in EPA's *Report on the Environment*, available at: www.epa.gov/roe.

References

Vose, R.S., and M.J. Menne. 2004. A method to determine station density requirements for climate observing networks. J. Climate 17(15):2961–2971.

Heavy Precipitation

Identification

1. Indicator Description

This indicator tracks the frequency of heavy precipitation events in the United States between 1895 and 2013. Heavy precipitation is a useful indicator because climate change can affect the intensity and frequency of precipitation. Warmer oceans increase the amount of water that evaporates into the air, and when more moisture-laden air moves over land or converges into a storm system, it can produce more intense precipitation—for example, heavier rain and snow storms (Tebaldi et al., 2006). The potential impacts of heavy precipitation include crop damage, soil erosion, flooding, and diminished water quality.

Components of this indicator include:

- Percent of land area in the contiguous 48 states experiencing abnormal amounts of annual rainfall from one-day precipitation events (Figure 1).
- Percent of land area in the contiguous 48 states with unusually high annual precipitation (Figure 2).

2. Revision History

April 2010:	Indicator posted.
December 2011:	Updated with data through 2010.
March 2012:	Updated with data through 2011.
August 2013:	Updated indicator on EPA's website with data through 2012.
March 2014:	Updated with data through 2013.

Data Sources

3. Data Sources

This indicator is based on precipitation measurements collected at weather stations throughout the contiguous 48 states. Most of the stations are part of the U.S. Historical Climatology Network (USHCN), a database compiled and managed by the National Oceanic and Atmospheric Administration's (NOAA's) National Climatic Data Center (NCDC). Indicator data were obtained from NCDC.

4. Data Availability

USHCN precipitation data are maintained at NOAA's NCDC, and the data are distributed on various computer media (e.g., anonymous FTP sites), with no confidentiality issues limiting accessibility. Users can link to the data online at: www.ncdc.noaa.gov/oa/climate/research/ushcn/#access.

Appropriate metadata and "readme" files are appended to the data so that they are discernible for analysis. For example, see: ftp://ftp.ncdc.noaa.gov/pub/data/ushcn/v2/monthly/readme.txt.

Figure 1. Extreme One-Day Precipitation Events in the Contiguous 48 States, 1910–2013

NOAA has calculated each of the components of the U.S. Climate Extremes Index (CEI) and has made these data files publicly available. The data set for extreme precipitation (CEI step 4) can be downloaded from: ftp://ftp.ncdc.noaa.gov/pub/data/cei/dk-step4.01-12.results. A "readme" file (at ftp://ftp.ncdc.noaa.gov/pub/data/cei) explains the contents of the data files.

Figure 2. Unusually High Annual Precipitation in the Contiguous 48 States, 1895–2013

Standardized Precipitation Index (SPI) data are publicly available and can be downloaded from: ftp://ftp.ncdc.noaa.gov/pub/data/cirs. This indicator uses 12-month SPI data, which are found in the file "drd964x.sp12.txt." This FTP site also includes a "readme" file that explains the contents of the data files.

Constructing Figure 2 required additional information about the U.S. climate divisions. The land area of each climate division can be found by going to: www.ncdc.noaa.gov/oa/climate/surfaceinventories.html and viewing the "U.S. climate divisions" file (exact link: ftp://ftp.ncdc.noaa.gov/pub/data/inventories/DIV-AREA.TXT). For a guide to the numerical codes assigned to each state, see: ftp://ftp.ncdc.noaa.gov/pub/data/inventories/COOP-STATE-CODES.TXT.

Methodology

5. Data Collection

This indicator is based on precipitation measurements collected by a network of thousands of weather stations spread throughout the contiguous 48 states. These stations are currently overseen by NOAA, and they use standard gauges to measure the amount of precipitation received on a daily basis. Some of the stations in the USHCN are first-order weather stations, but the majority are selected from approximately 5,000 cooperative weather stations in the United States.

NOAA's NCDC has published extensive documentation about data collection methods for the USHCN data set. See: www.ncdc.noaa.gov/oa/climate/research/ushcn, which lists a set of technical reports and peer-reviewed articles that provide more detailed information about USHCN methodology. See: www.ncdc.noaa.gov/oa/ncdc.html for information on other types of weather stations that have been used to supplement the USHCN record.

6. Indicator Derivation

Heavy precipitation can be examined in many different ways. For example, the prevalence of extreme individual events can be characterized in terms of the number of 24-hour events exceeding a fixed precipitation threshold (e.g., 1 or 2 inches), the number of 24-hour events considered "extreme" based on the historical distribution of precipitation events at a given location (i.e., a percentile-based approach), the proportion of annual precipitation derived from "extreme" 24-hour events, or other approaches. This indicator uses a percentile-based approach in Figure 1 because it accounts for regional differences in what might be considered "heavy" precipitation (e.g., 1 inch in a day might be common in some places but not in others) and because the data are readily available as part of NOAA's CEI. Figure 2 complements this analysis by considering total annual precipitation, which reflects the cumulative influence of heavy precipitation events occurring throughout the year.

Figure 1 and Figure 2 are based on similar raw data (i.e., daily precipitation measurements), but were developed using two different models because they show trends in extreme precipitation from two different perspectives.

Figure 1. Extreme One-Day Precipitation Events in the Contiguous 48 States, 1910–2013

Figure 1 was developed as part of NOAA's CEI, an index that uses six different variables to examine trends in extreme weather and climate. This figure shows trends in the prevalence of extreme one-day precipitation events, based on a component of NOAA's CEI (labeled as Step 4) that looks at the percentage of land area within the contiguous 48 states that experienced a much greater than normal proportion of precipitation derived from extreme one-day precipitation events in any given year.

In compiling the CEI, NOAA applied more stringent criteria to select only those stations with data for at least 90 percent of the days in each year, as well as 90 percent of the days during the full period of record. Applying these criteria resulted in the selection of a subset of USHCN stations. To supplement the USHCN record, the CEI (and hence Figure 1) also includes data from NOAA's Cooperative Summary of the Day (TD3200) and pre-1948 (TD3206) daily precipitation stations. This resulted in a total of over 1,300 precipitation stations.

NOAA scientists computed the data for the CEI and calculated the percentage of land area for each year. They performed these steps by dividing the contiguous 48 states into a 1-degree by 1-degree grid and using data from one station in each grid box, rather than multiple stations. This was done to eliminate many of the artificial extremes that resulted from a changing number of available stations over time.

For each grid cell, the indicator looks at what portion of the total annual precipitation occurred on days that had extreme precipitation totals. Thus, the indicator essentially describes what percentage of precipitation is arriving in short, intense bursts. "Extreme" is defined as the highest 10th percentile, meaning an extreme one-day event is one in which the total precipitation received at a given location during the course of the day is at the upper end of the distribution of expected values (i.e., the distribution of all one-day precipitation totals at that location during the period of record). After extreme one-day events were identified, the percentage of annual precipitation occurring on extreme days was calculated for each year at each location. The subsequent step looked at the distribution of these percentage values over the full period of record, then identified all years that were in the highest 10th percentile. These years were considered to have a "greater than normal" amount of precipitation derived from extreme precipitation events at a given location. The top 10th percentile was chosen so as to give the overall index an expected value of 10 percent. Finally, data were aggregated nationwide to determine the percentage of land area with greater than normal precipitation derived from extreme events in each year.

The CEI can be calculated for individual seasons or for an entire year. This indicator uses the annual CEI, which is shown by the columns in Figure 1. To smooth out some of the year-to-year variability, EPA applied a nine-point binomial filter, which is plotted at the center of each nine-year window. For example, the smoothed value from 2002 to 2010 is plotted at year 2006. NOAA NCDC recommends this approach and has used it in the official online reporting tool for the CEI.

EPA used endpoint padding to extend the nine-year smoothed lines all the way to the ends of the period of record. As recommended by NCDC, EPA calculated smoothed values as follows: if 2013 was the most recent year with data available, EPA calculated smoothed values to be centered at 2010, 2011, 2012, and 2013 by inserting the 2013 data point into the equation in place of the as-yet-unreported annual data points for 2014 and beyond. EPA used an equivalent approach at the beginning of the time series.

The CEI has been extensively documented and refined over time to provide the best possible representation of trends in extreme weather and climate. For an overview of how NOAA constructed Step 4 of the CEI, see: www.ncdc.noaa.gov/extremes/cei. This page provides a list of references that describe analytical methods in greater detail. In particular, see Gleason et al. (2008).

Figure 2. Unusually High Annual Precipitation in the Contiguous 48 States, 1895–2013

Figure 2 shows trends in the occurrence of abnormally high annual total precipitation based on the SPI, which is an index based on the probability of receiving a particular amount of precipitation in a given location. Thus, this index essentially compares the actual amount of annual precipitation received at a particular location with the amount that would be expected based on historical records. An SPI value of zero represents the median of the historical distribution; a negative SPI value represents a drier-than-normal period and a positive value represents a wetter-than-normal period.

The Western Regional Climate Center (WRCC) calculates the SPI by dividing the contiguous 48 states into 344 regions called climate divisions and analyzing data from weather stations within each division. A typical division has 10 to 50 stations, some from USHCN and others from the broader set of cooperative weather stations. For a given time period, WRCC calculated a single SPI value for each climate division based on an unweighted average of data from all stations within the division. This procedure has been followed for data from 1931 to present. A regression technique was used to compute divisional values prior to 1931 (Guttman and Quayle, 1996).

WRCC and NOAA calculate the SPI for various time periods ranging from one month to 24 months. This indicator uses the 12-month SPI data reported for the end of December of each year (1895 to 2013). The 12-month SPI is based on precipitation totals for the previous 12 months, so a December 12-month SPI value represents conditions over the full calendar year.

To create Figure 2, EPA identified all climate divisions with an SPI value of +2.0 or greater in a given year, where +2.0 is a suggested threshold for "abnormally high" precipitation (i.e., the upper tail of the historical distribution). For each year, EPA then determined what percentage of the total land area of the contiguous 48 states these "abnormally high" climate divisions represent. This annual percentage value is represented by the thin curve in the graph. To smooth out some of the year-to-year variability, EPA applied a nine-point binomial filter, which is plotted at the center of each nine-year window. For example, the smoothed value from 2002 to 2010 is plotted at year 2006. NOAA NCDC recommends this approach and has used it in the official online reporting tool for the CEI (the source of Figure 1).

EPA used endpoint padding to extend the nine-year smoothed lines all the way to the ends of the period of record. As recommended by NCDC, EPA calculated smoothed values as follows: If 2013 was the most recent year with data available, EPA calculated smoothed values to be centered at 2010, 2011, 2012, and 2013 by inserting the 2013 data point into the equation in place of the as-yet-unreported annual data points for 2014 and beyond. EPA used an equivalent approach at the beginning of the time series.

Like the CEI, the SPI is extensively documented in the peer-reviewed literature. The SPI is particularly useful with drought and precipitation indices because it can be applied over a variety of time frames and because it allows comparison of different locations and different seasons on a standard scale.

For an overview of the SPI and a list of resources describing methods used in constructing this index, see NDMC (2011) and the following websites: http://lwf.ncdc.noaa.gov/oa/climate/research/prelim/drought/spi.html and www.wrcc.dri.edu/spi/explanation.html. For more information on climate divisions and the averaging and regression processes used to generalize values within each division, see Guttman and Quayle (1996).

General Discussion

This indicator does not attempt to project data backward before the start of regular data collection or forward into the future. All values of the indicator are based on actual measured data. No attempt has been made to interpolate days with missing data. Rather, the issue of missing data was addressed in the site selection process by including only those stations that had very few missing data points.

7. Quality Assurance and Quality Control

USHCN precipitation data have undergone extensive quality assurance and quality control (QA/QC) procedures to identify errors and biases in the data and either remove these stations from the time series or apply correction factors. These quality control procedures are summarized at: www.ncdc.noaa.gov/oa/climate/research/ushcn/#processing. A series of data corrections was developed to address specific potential problems in trend estimation in USHCN Version 2. They include:

- Removal of duplicate records.
- Procedures to deal with missing data.
- Testing and correcting for artificial discontinuities in a local station record, which might reflect station relocation or instrumentation changes.

Data from weather stations also undergo routine QC checks before they are added to historical databases in their final form. These steps are typically performed within four months of data collection (NDMC, 2011).

QA/QC procedures are not readily available for the CEI and SPI, but both of these indices have been published in the peer-reviewed literature, indicating a certain degree of rigor.

Analysis

8. Comparability Over Time and Space

To be included in the USHCN, a station had to meet certain criteria for record longevity, data availability (percentage of missing values), spatial coverage, and consistency of location (i.e., experiencing few station changes). The period of record varies for each station but generally includes most of the 20th century. One of the objectives in establishing the USHCN was to detect secular changes in regional rather than local climate. Therefore, stations included in the network are only those believed to not be influenced to any substantial degree by artificial changes of local environments.

9. Data Limitations

Factors that may impact the confidence, application, or conclusions drawn from this indicator are as follows:

1. Both figures are national in scope, meaning they do not provide information about trends in extreme or heavy precipitation on a local or regional scale.

2. Weather monitoring stations tend to be closer together in the eastern and central states than in the western states. In areas with fewer monitoring stations, heavy precipitation indicators are less likely to reflect local conditions accurately.

3. The indicator does not include Alaska, which has seen some notable changes in heavy precipitation in recent years (e.g., Gleason et al., 2008).

10. Sources of Uncertainty

Error estimates are not readily available for daily precipitation measurements or for the CEI and SPI calculations that appear in this indicator. In general, uncertainties in precipitation data increase as one goes back in time, as there are fewer stations early in the record. However, these uncertainties should not be sufficient to undermine the fundamental trends in the data. The USHCN has undergone extensive testing to identify errors and biases in the data and either remove these stations from the time series or apply scientifically appropriate correction factors to improve the utility of the data. In addition, both parts of the indicator have been restricted to stations meeting specific criteria for data availability.

11. Sources of Variability

Precipitation varies from location to location and from year to year as a result of normal variation in weather patterns, multi-year climate cycles such as the El Niño–Southern Oscillation and Pacific Decadal Oscillation, and other factors. This indicator accounts for these factors by presenting a long-term record (a century of data) and aggregating consistently over time and space.

12. Statistical/Trend Analysis

EPA has determined that the time series in Figure 1 has an increasing trend of approximately half a percentage point per decade and the time series in Figure 2 has an increasing trend of approximately 0.15 percentage points per decade. Both of these trends were calculated by ordinary least-squares regression, which is a common statistical technique for identifying a first-order trend. Analyzing the significance of these trends would potentially require consideration of serial correlation and other more complex statistical factors.

References

Gleason, K.L., J.H. Lawrimore, D.H. Levinson, T.R. Karl, and D.J. Karoly. 2008. A revised U.S. climate extremes index. J. Climate 21:2124–2137.

Guttman, N.B., and R.G. Quayle. 1996. A historical perspective of U.S. climate divisions. Bull. Am. Meteorol. Soc. 77(2):293–303.

NDMC (National Drought Mitigation Center). 2011. Data source and methods used to compute the Standardized Precipitation Index.
http://drought.unl.edu/MonitoringTools/ClimateDivisionSPI/DataSourceMethods.aspx.

Tebaldi, C., K. Hayhoe, J.M. Arblaster, and G.A. Meehl. 2006. Going to the extremes: An intercomparison of model-simulated historical and future changes in extreme events. Climatic Change 79:185–211.

Drought

Identification

1. Indicator Description

This indicator measures drought conditions in the United States from 1895 to 2013. Drought can affect agriculture, water supplies, energy production, and many other aspects of society. Drought relates to climate change because rising average temperatures alter the Earth's water cycle, increasing the overall rate of evaporation. An increase in evaporation makes more water available in the air for precipitation, but contributes to drying over some land areas, leaving less moisture in the soil. As the climate continues to change, many areas are likely to experience increased precipitation and increased risk of flooding, while areas far from storm tracks are likely to experience less precipitation and increased risk of drought.

Components of this indicator include:

- Average drought conditions in the contiguous 48 states over time, based on the Palmer Drought Severity Index (Figure 1).

- Percent of U.S. lands classified under drought conditions in recent years, based on an index called the U.S. Drought Monitor (Figure 2).

2. Revision History

April 2010:	Indicator posted.
December 2011:	Updated with U.S. Drought Monitor data through 2010; added a new figure based on the Palmer Drought Severity Index (PDSI).
January 2012:	Updated with data through 2011.
August 2013:	Updated indicator on EPA's website with data through 2012.
March 2014:	Updated with data through 2013.

Data Sources

3. Data Sources

Data for Figure 1 were obtained from the National Oceanic and Atmospheric Administration's (NOAA's) National Climatic Data Center (NCDC), which maintains a large collection of climate data online.

Data for Figure 2 were provided by the U.S. Drought Monitor, which maintains current and archived data at: http://droughtmonitor.unl.edu.

4. Data Availability

Figure 1. Average Drought Conditions in the Contiguous 48 States, 1895–2013

NCDC provides access to monthly values of the PDSI averaged across the entire contiguous 48 states, which EPA downloaded for this indicator. These data are available at: www7.ncdc.noaa.gov/CDO/CDODivisionalSelect.jsp. This website also provides access to monthly PDSI values for nine broad regions, individual states, and 344 smaller regions called climate divisions (each state has one to 10 climate divisions). For accompanying metadata, see: www7.ncdc.noaa.gov/CDO/DIV_DESC.txt.

PDSI values are calculated from precipitation and temperature measurements collected by weather stations within each climate division. Individual station measurements and metadata are available through NCDC's website (www.ncdc.noaa.gov/oa/ncdc.html).

Figure 2. U.S. Lands Under Drought Conditions, 2000–2013

U.S. Drought Monitor data can be obtained from: http://droughtmonitor.unl.edu/MapsAndData/DataTables.aspx. Select "United States" to view the historical data that were used for this indicator. For each week, the data table shows what percentage of land area was under the following drought conditions:

1. None
2. D0–D4
3. D1–D4
4. D2–D4
5. D3–D4
6. D4 alone

This indicator covers the time period from 2000 to 2013. Although data were available for parts of 1999 and 2014 at the time EPA last updated this indicator, EPA chose to report only full years.

Drought Monitor data are based on a wide variety of underlying sources. Some are readily available from public websites; others might require specific database queries or assistance from the agencies that collect and/or compile the data. For links to many of the data sources, see: http://droughtmonitor.unl.edu/SupplementalInfo/Links.aspx.

Methodology

5. Data Collection

Figure 1. Average Drought Conditions in the Contiguous 48 States, 1895–2013

The PDSI is calculated from daily temperature measurements and precipitation totals collected at thousands of weather stations throughout the United States. These stations are overseen by NOAA, and they use standard instruments to measure temperature and precipitation. Some of these stations are first-order stations operated by NOAA's National Weather Service. The remainder are Cooperative

Observer Program (COOP) stations operated by other organizations using trained observers and equipment and procedures prescribed by NOAA. For an inventory of U.S. weather stations and information about data collection methods, see: www.ncdc.noaa.gov/oa/land.html#dandp, www.ncdc.noaa.gov/oa/climate/research/ushcn, and the technical reports and peer-reviewed papers cited therein.

Figure 2. U.S. Lands Under Drought Conditions, 2000–2013

Figure 2 is based on the U.S. Drought Monitor, which uses a comprehensive definition of drought that accounts for a large number of different physical variables. Many of the underlying variables reflect weather and climate, including daily precipitation totals collected at weather stations throughout the United States, as described above for Figure 1. Other parameters include measurements of soil moisture, streamflow, reservoir and groundwater levels, and vegetation health. These measurements are generally collected by government agencies following standard methods, such as a national network of stream gauges that measure daily and weekly flows, comprehensive satellite mapping programs, and other systematic monitoring networks. Each program has its own sampling or monitoring design. The Drought Monitor and the other drought indices that contribute to it have been formulated to rely on measurements that offer sufficient temporal and spatial resolution.

The U.S. Drought Monitor has five primary inputs:

- The PDSI.
- The Soil Moisture Model, from NOAA's Climate Prediction Center.
- Weekly streamflow data from the U.S. Geological Survey.
- The Standardized Precipitation Index (SPI), compiled by NOAA and the Western Regional Climate Center (WRCC).
- A blend of objective short- and long-term drought indicators (short-term drought indicator blends focus on 1- to 3-month precipitation totals; long-term blends focus on 6 to 60 months).

At certain times and in certain locations, the Drought Monitor also incorporates one or more of the following additional indices, some of which are particularly well-suited to the growing season and others of which are ideal for snowy areas or ideal for the arid West:

- A topsoil moisture index from the U.S. Department of Agriculture's National Agricultural Statistics Service.
- The Keetch-Byram Drought Index.
- Vegetation health indices based on satellite imagery from NOAA's National Environmental Satellite, Data, and Information Service (NESDIS).
- Snow water content.
- River basin precipitation.
- The Surface Water Supply Index (SWSI).
- Groundwater levels.
- Reservoir storage.
- Pasture or range conditions.

For more information on the other drought indices that contribute to the Drought Monitor, including the data used as inputs to these other indices, see: http://drought.unl.edu/Planning/Monitoring/ComparisonofIndicesIntro.aspx.

To find information on underlying sampling methods and procedures for constructing some of the component indices that go into determining the U.S. Drought Monitor, one will need to consult a variety of additional sources. For example, as described above for Figure 1, NCDC has published extensive documentation about methods for collecting precipitation data.

6.   Indicator Derivation

Figure 1. Average Drought Conditions in the Contiguous 48 States, 1895–2013

PDSI calculations are designed to reflect the amount of moisture available at a particular place and time, based on the amount of precipitation received as well as the temperature, which influences evaporation rates. The formula for creating this index was originally proposed in the 1960s (Palmer, 1965). Since then, the methods have been tested extensively and used to support hundreds of published studies. The PDSI is the most widespread and scientifically vetted drought index in use today.

The PDSI was designed to characterize long-term drought (i.e., patterns lasting a month or more). Because drought is cumulative, the formula takes precipitation and temperature data from previous weeks and months into account. Thus, a single rainy day is unlikely to cause a dramatic shift in the index.

PDSI values are normalized relative to long-term average conditions at each location, which means this method can be applied to any location regardless of how wet or dry it typically is. NOAA currently uses 1931–1990 as its long-term baseline. The index essentially measures deviation from normal conditions. The PDSI takes the form of a numerical value, generally ranging from -6 to +6. A value of zero reflects average conditions. Negative values indicate drier-than-average conditions and positive values indicate wetter-than-average conditions. NOAA provides the following interpretations for specific ranges of the index:

- 0 to -0.5 = normal
- -0.5 to -1.0 = incipient drought
- -1.0 to -2.0 = mild drought
- -2.0 to -3.0 = moderate drought
- -3.0 to -4.0 = severe drought
- < -4.0 = extreme drought

Similar adjectives can be applied to positive (wet) values.

NOAA calculates monthly values of the PDSI for each of the 344 climate divisions within the contiguous 48 states. These values are calculated from weather stations reporting both temperature and precipitation. All stations within a division are given equal weight. NOAA also combines PDSI values from all climate divisions to derive a national average for every month.

EPA obtained monthly national PDSI values from NOAA, then calculated annual averages. To smooth out some of the year-to-year variability, EPA applied a nine-point binomial filter, which is plotted at the center of each nine-year window. For example, the smoothed value from 2002 to 2010 is plotted at year

2006. NOAA NCDC recommends this approach. Figure 1 shows both the annual values and the smoothed curve.

EPA used endpoint padding to extend the nine-year smoothed lines all the way to the ends of the period of record. As recommended by NCDC, EPA calculated smoothed values as follows: if 2013 was the most recent year with data available, EPA calculated smoothed values to be centered at 2010, 2011, 2012, and 2013 by inserting the 2013 data point into the equation in place of the as-yet-unreported annual data points for 2014 and beyond. EPA used an equivalent approach at the beginning of the time series.

For more information about NOAA's processing methods, see the metadata file at: www7.ncdc.noaa.gov/CDO/DIV_DESC.txt. NOAA's website provides a variety of other references regarding the PDSI at: www.ncdc.noaa.gov/oa/climate/research/prelim/drought/palmer.html.

In March 2013, NOAA corrected minor errors in the computer code used to process soil moisture values, which feed into the computation of the PDSI. This change caused slight revisions to historical data compared with what EPA presented in Figure 1 prior to August 2013. Although most data were not substantially changed, minor but discernible differences appeared in data after 2005. NOAA discusses these improvements in full at: www.ncdc.noaa.gov/sotc/national/2013/3/supplemental/page-7.

Figure 2. U.S. Lands Under Drought Conditions, 2000–2013

The National Drought Mitigation Center at the University of Nebraska–Lincoln produces the U.S. Drought Monitor with assistance from many other climate and water experts at the federal, regional, state, and local levels. For each week, the Drought Monitor labels areas of the country according to the intensity of any drought conditions present. An area experiencing drought is assigned a score ranging from D0, the least severe drought, to D4, the most severe. For definitions of these classifications, see: http://droughtmonitor.unl.edu/AboutUs/ClassificationScheme.aspx.

Drought Monitor values are determined from the five major components and other supplementary factors listed in Section 5. A table on the Drought Monitor website (http://droughtmonitor.unl.edu/AboutUs/ClassificationScheme.aspx) explains the range of observed values for each major component that would result in a particular Drought Monitor score. The final index score is based to some degree on expert judgment, however. For example, expert analysts resolve discrepancies in cases where the five major components might not coincide with one another. They might assign a final Drought Monitor score based on what the majority of the components suggest, or they might weight the components differently according to how well they perform in various parts of the country and at different times of the year. Experts also determine what additional factors to consider for a given time and place and how heavily to weight these supplemental factors. For example, snowpack is particularly important in the West, where it has a strong bearing on water supplies.

From the Drought Monitor's public website, EPA obtained data covering the contiguous 48 states plus Alaska, Hawaii, and Puerto Rico, then performed a few additional calculations. The original data set reports cumulative categories (for example, "D2–D4" and "D3–D4"), so EPA had to subtract one category from another in order to find the percentage of land area belonging to each individual drought category (e.g., D2 alone). EPA also calculated annual averages to support some of the statements presented in the "Key Points" for this indicator.

No attempt has been made to portray data outside the time and space in which measurements were made. Measurements are collected at least weekly (in the case of some variables like precipitation and streamflow, at least daily) and used to derive weekly maps for the U.S. Drought Monitor. Values are generalized over space by weighting the different factors that go into calculating the overall index and applying expert judgment to derive the final weekly map and the corresponding totals for affected area.

For more information about how the Drought Monitor is calculated, including percentiles associated with the occurrence of each of the D0–D4 classifications, see Svoboda et al. (2002), along with the documentation provided on the Drought Monitor website at: http://droughtmonitor.unl.edu.

7. Quality Assurance and Quality Control

Figure 1. Average Drought Conditions in the Contiguous 48 States, 1895–2013

Data from weather stations go through a variety of quality assurance and quality control (QA/QC) procedures before they can be added to historical databases in their final form. NOAA's U.S. Historical Climatology Network—one of the main weather station databases—follows strict QA/QC procedures to identify errors and biases in the data and then either remove these stations from the time series or apply correction factors. Procedures for the USHCN are summarized at: www.ncdc.noaa.gov/oa/climate/research/ushcn/#processing. Specific to this indicator, NOAA's metadata file (www7.ncdc.noaa.gov/CDO/DIV_DESC.txt) and Karl et al. (1986) describe steps that have been taken to reduce biases associated with differences in the time of day when temperature observations are reported.

Figure 2. U.S. Lands Under Drought Conditions, 2000–2013

QA/QC procedures for the overall U.S. Drought Monitor data set are not readily available. Each underlying data source has its own methodology, which typically includes some degree of QA/QC. For example, precipitation and temperature data are verified and corrected as described above for Figure 1. Some of the other underlying data sources have QA/QC procedures available online, but others do not.

Analysis

8. Comparability Over Time and Space

Figure 1. Average Drought Conditions in the Contiguous 48 States, 1895–2013

PDSI calculation methods have been applied consistently over time and space. In all cases, the index relies on the same underlying measurements (precipitation and temperature). Although fewer stations were collecting weather data during the first few decades of the analysis, NOAA has determined that enough stations were available starting in 1895 to calculate valid index values for the contiguous 48 states as a whole.

Figure 2. U.S. Lands Under Drought Conditions, 2000–2013

The resolution of the U.S. Drought Monitor has improved over time. When the Drought Monitor began to be calculated in 1999, many of the component indicators used to determine drought conditions were reported at the climate division level. Many of these component indicators now include data from the county and sub-county level. This change in resolution over time can be seen in the methods used to draw contour lines on Drought Monitor maps.

The drought classification scheme used for this indicator is produced by combining data from several different sources. Different locations may use different primary sources—or the same sources, weighted differently. These data are combined to reflect the collective judgment of experts and in some cases are adjusted to reconcile conflicting trends shown by different data sources over different time periods.

Though data resolution and mapping procedures have varied somewhat over time and space, the fundamental construction of the indicator has remained consistent.

9. Data Limitations

Factors that may impact the confidence, application, or conclusions drawn from this indicator are as follows:

1. The indicator gives a broad overview of drought conditions in the United States. It is not intended to replace local or state information that might describe conditions more precisely for a particular region. Local or state entities might monitor different variables to meet specific needs or to address local problems. As a consequence, there could be water shortages or crop failures within an area not designated as a drought area, just as there could be locations with adequate water supplies in an area designated as D3 or D4 (extreme or exceptional) drought.

2. Because this indicator focuses on national trends, it does not show how drought conditions vary by region. For example, even if half of the country suffered from severe drought, Figure 1 could show an average index value close to zero if the rest of the country was wetter than average. Thus, Figure 1 might understate the degree to which droughts are becoming more severe in some areas, while other places receive more rain as a result of climate change.

3. Although the PDSI is arguably the most widely used drought index, it has some limitations that have been documented extensively in the literature. While the use of just two variables (precipitation and temperature) makes this index relatively easy to calculate over time and space, drought can have many other dimensions that these two variables do not fully capture. For example, the PDSI loses accuracy in areas where a substantial portion of the water supply comes from snowpack.

4. Indices such as the U.S. Drought Monitor seek to address the limitations of the PDSI by incorporating many more variables. However, the Drought Monitor is relatively new and cannot yet be used to assess long-term climate trends.

5. The drought classification scheme used for Figure 2 is produced by combining data from several different sources. These data are combined to reflect the collective judgment of experts and in some cases are adjusted to reconcile conflicting trends shown by different data sources over different time periods.

10. Sources of Uncertainty

Error estimates are not readily available for national average PDSI, the U.S. Drought Monitor, or the underlying measurements that contribute to this indicator. It is not clear how much uncertainty might be associated with the component indices that go into formulating the Drought Monitor or the process of compiling these indices into a single set of weekly values through averaging, weighting, and expert judgment.

11. Sources of Variability

Conditions associated with drought naturally vary from place to place and from one day to the next, depending on weather patterns and other factors. Both figures address spatial variability by presenting aggregate national trends. Figure 1 addresses temporal variability by using an index that is designed to measure long-term drought and is not easily swayed by short-term conditions. Figure 1 also provides an annual average, along with a nine-year smoothed average. Figure 2 smoothes out some of the inherent variability in drought measurement by relying on many indices, including several with a long-term focus. While Figure 2 shows noticeable week-to-week variability, it also reveals larger year-to-year patterns.

12. Statistical/Trend Analysis

This indicator does not report on the slope of the trend in PDSI values over time, nor does it calculate the statistical significance of this trend.

Because data from the U.S. Drought Monitor are only available for the most recent decade, this metric is too short-lived to be used for assessing long-term climate trends. Furthermore, there is no clear long-term trend in Figure 2. With continued data collection, future versions of this indicator should be able to paint a more statistically robust picture of long-term trends in Drought Monitor values.

References

Karl, T.R., C.N. Williams, Jr., P.J. Young, and W.M. Wendland. 1986. A model to estimate the time of observation bias associated with monthly mean maximum, minimum, and mean temperatures for the Unites States. J. Clim. Appl. Meteorol. 25(1):145–160.

Palmer, W.C. 1965. Meteorological drought. Res. Paper No.45. Washington, DC: U.S. Department of Commerce.

Svoboda, M., D. Lecomte, M. Hayes, R. Heim, K. Gleason, J. Angel, B. Rippey, R. Tinker, M. Palecki, D. Stooksbury, D. Miskus, and S. Stephens. 2002. The drought monitor. B. Am. Meteorol. Soc. 83(8):1181–1190.

Temperature and Drought in the Southwest

Identification

1. Description

This regional feature measures trends in drought conditions and temperature in six states: Arizona, California, Colorado, Nevada, New Mexico, and Utah. The metrics presented in this feature provide insight into how climate change is affecting areas of vulnerability in the U.S. Southwest. The Southwest is particularly vulnerable to the effects of drought because water is already scarce in this region and because the region is particularly dependent on surface water supplies like Lake Mead, which are vulnerable to evaporation. As described in the U.S. and Global Temperature indicator and the Drought indicator, climate change can result in changes in temperature and drought conditions.

Components of this regional feature include:

- Spatial and temporal trends in temperature anomalies from 1895 to 2013 (Figure 1).

- Percent of lands classified under drought conditions in recent years, based on the U.S. Drought Monitor Drought Severity Classification system (Figure 2).

- Spatial and temporal trends in drought severity from 1895 to 2013, based on the Palmer Drought Severity Index (PDSI) (Figure 3).

2. Revision History

December 2013:   Feature proposed.
January 2014:   Updated Figure 2 with 2013 data.
March 2014:   Updated Figures 1 and 3 with 2013 data.

Data Sources

3. Data Sources

Data for Figures 1 and 3 were obtained from the National Oceanic and Atmospheric Administration's (NOAA's) National Climatic Data Center (NCDC). This data set provides information on average temperatures, precipitation, and several comparative measures of precipitation (e.g., Standardized Precipitation Index) and drought severity (e.g., Palmer's Drought Severity Index). Data have been compiled for individual climate divisions (each state has up to 10 climate divisions; see: www.ncdc.noaa.gov/monitoring-references/maps/us-climate-divisions.php) and are available from 1895 to present.

Data for Figure 2 were provided by the U.S. Drought Monitor, which maintains current and archived data at: http://droughtmonitor.unl.edu.

4. Data Availability

Data for Figures 1 and 3 are derived from tabular data available through the NCDC online data portal, Climate Data Online (CDO), at divisional, state, regional, and national scales from 1895 through present. The entire data set is available at: www7.ncdc.noaa.gov/CDO/CDODivisionalSelect.jsp.

U.S. Drought Monitor data for Figure 2 can be obtained from: http://droughtmonitor.unl.edu/MapsAndData/DataTables.aspx. For each week, the data table shows what percentage of land area was under drought conditions D0 (abnormally dry) through D4 (exceptional drought). This component of the regional feature covers the time period from 2000 to 2013. Although data were available for part of 2014 at the time EPA last updated this feature, EPA chose to report only full years.

Drought Monitor data are based on a wide variety of underlying sources. Some are readily available from public websites; others might require specific database queries or assistance from the agencies that collect and/or compile the data. For links to many of the data sources, see: http://droughtmonitor.unl.edu/SupplementalInfo/Links.aspx.

Methodology

5. Data Collection

Figure 1. Average Temperatures in the Southwestern United States, 2000–2013 Versus Long-Term Average

This figure was developed by analyzing temperature records from thousands of weather stations that constitute NCDC's *n*ClimDiv data set. These stations are overseen by NOAA, and they use standard instruments to measure temperature and precipitation. Some of these stations are first-order stations operated by NOAA's National Weather Service (NWS). The remainder are Cooperative Observer Program (COOP) stations operated by other organizations using trained observers and equipment and procedures prescribed by NOAA. These stations generally measure temperature at least hourly, and they record the minimum temperature for each 24-hour time span. Cooperative observers include state universities, state and federal agencies, and private individuals whose stations are managed and maintained by NWS. Observers are trained to collect data, and the NWS provides and maintains standard equipment to gather these data. The NWS/COOP data set represents the core climate network of the United States (Kunkel et al., 2005). Data collected by these sites are referred to as U.S. Daily Surface Data or Summary of the Day data.

Altogether, the six states covered in this feature are home to more than 6,100 past and present NWS and COOP stations. For a complete inventory of U.S. weather stations and information about data collection methods, see: www.ncdc.noaa.gov/oa/land.html#dandp, www.ncdc.noaa.gov/oa/climate/research/ushcn, and the technical reports and peer-reviewed papers cited therein. Additional information about the NWS COOP data set is available at: www.nws.noaa.gov/os/coop/what-is-coop.html. Sampling procedures are described in Kunkel et al. (2005) and in the full metadata for the COOP data set available at: www.nws.noaa.gov/om/coop.

Figure 2. Southwestern U.S. Lands Under Drought Conditions, 2000–2013

Figure 2 is based on the U.S. Drought Monitor, which uses a comprehensive definition of drought that accounts for a large number of different physical variables. Many of the underlying variables reflect weather and climate, including daily precipitation totals collected at thousands of weather stations, as described for Figures 1 and 3. Other parameters include measurements of soil moisture, streamflow, reservoir and groundwater levels, and vegetation health. These measurements are generally collected by government agencies following standard methods, such as a national network of stream gauges that measure daily (and weekly) flows, comprehensive satellite mapping programs, and other systematic monitoring networks. Each program has its own sampling or monitoring design. The Drought Monitor and the other drought indices that contribute to it have been formulated such that they rely on measurements that offer sufficient temporal and spatial resolution.

The U.S. Drought Monitor has five primary inputs:

- The PDSI.
- The Soil Moisture Model, from NOAA's Climate Prediction Center.
- Weekly streamflow data from the U.S. Geological Survey.
- The Standardized Precipitation Index (SPI), compiled by NOAA and the Western Regional Climate Center (WRCC).
- A blend of objective short- and long-term drought indicators (short-term drought indicator blends focus on 1- to 3-month precipitation totals; long-term blends focus on 6 to 60 months).

At certain times and in certain locations, the Drought Monitor also incorporates one or more of the following additional indices, some of which are particularly well-suited to the growing season and others of which are ideal for snowy areas or ideal for the arid West:

- A topsoil moisture index from the U.S. Department of Agriculture's National Agricultural Statistics Service.
- The Keetch-Byram Drought Index.
- Vegetation health indices based on satellite imagery from NOAA's National Environmental Satellite, Data, and Information Service (NESDIS).
- Snow water content.
- River basin precipitation.
- The Surface Water Supply Index (SWSI).
- Groundwater levels.
- Reservoir storage.
- Pasture or range conditions.

For more information on the other drought indices that contribute to the Drought Monitor, including the data used as inputs to these other indices, see:
http://drought.unl.edu/Planning/Monitoring/ComparisonofIndicesIntro.aspx.

To find information on underlying sampling methods and procedures for constructing some of the component indices that go into determining the U.S. Drought Monitor, one will need to consult a variety of additional sources. For example, as described for Figures 1 and 3, NCDC has published extensive documentation about methods for collecting precipitation data.

Figure 3. Drought Severity in the Southwestern United States, 1895–2013

The PDSI is calculated from daily temperature measurements and precipitation totals collected at thousands of weather stations, as described above for Figure 2. See the description for Figure 1 above for more information about these data collection networks.

6. Derivation

Figure 1. Average Temperatures in the Southwestern United States, 2000–2013 Versus Long-Term Average

NOAA used monthly mean temperatures at each weather station to calculate annual averages. Next, an annual average was determined for each climate division. To perform this step, NOAA used a grid-based computational approach known as climatologically-aided interpolation (Willmott and Robeson, 1995), which helps to address topographic variability. This technique is the hallmark of NOAA's *n*ClimDiv data product. Data from individual stations are combined in a grid with 5-kilometer resolution. To learn more about *n*ClimDiv, see: www.ncdc.noaa.gov/news/ncdc-introduces-national-temperature-index-page and: www.ncdc.noaa.gov/monitoring-references/maps/us-climate-divisions.php.

EPA calculated multi-year averages for each climate division, covering the full period of record (1895–2013) and the 21st century to date (2000–2013). The difference between the 21st century average and the 1895–2013 average is the anomaly shown in the Figure 1 map.

Figure 2. Southwestern U.S. Lands Under Drought Conditions, 2000–2013

The National Drought Mitigation Center at the University of Nebraska–Lincoln produces the U.S. Drought Monitor with assistance from many other climate and water experts at the federal, regional, state, and local levels. For each week, the Drought Monitor labels areas of the country according to the intensity of any drought conditions that may be present. An area experiencing drought is assigned a score ranging from D0, the least severe drought, to D4, the most severe. For definitions of these classifications, see: http://droughtmonitor.unl.edu/AboutUs/ClassificationScheme.aspx.

Drought Monitor values are determined from the five major components and other supplementary factors listed in Section 5. A table on the Drought Monitor website (http://droughtmonitor.unl.edu/AboutUs/ClassificationScheme.aspx) explains the range of observed values for each major component that would result in a particular Drought Monitor score. The final index score is based to some degree on expert judgment, however. For example, expert analysts resolve discrepancies in cases where the five major components might not coincide with one another. They might assign a final Drought Monitor score based on what the majority of the components suggest, or they might weight the components differently according to how well they perform in various parts of the country and at different times of the year. Experts also determine what additional factors to consider for a given time and place and how heavily to weight these supplemental factors. For example, snowpack is particularly important in the West, where it has a strong bearing on water supplies.

From the Drought Monitor's public website, EPA obtained weekly state-level Drought Monitor data for Arizona, California, Colorado, Nevada, New Mexico, and Utah. These data indicate the percentage of each state's area that falls into each of the Drought Monitor intensity classifications. To derive totals for the entire six-state region, EPA averaged the state-level data together for each week, weighted by state

area. This procedure used state areas as defined by the U.S. Census Bureau at:
https://www.census.gov/geo/reference/state-area.html.

No attempt has been made to portray data outside the time and space where measurements were made. Measurements are collected at least weekly (in the case of some variables like precipitation and streamflow, at least daily) and used to derive weekly maps for the U.S. Drought Monitor. Values are generalized over space by weighting the different factors that go into calculating the overall index and applying expert judgment to derive the final weekly map and the corresponding totals for affected area.

For more information about how the Drought Monitor is calculated, including percentiles associated with the occurrence of each of the D0–D4 classifications, see Svoboda et al. (2002) along with the documentation provided on the Drought Monitor website at: http://droughtmonitor.unl.edu.

Figure 3. Drought Severity in the Southwestern United States, 1895–2013

PDSI calculations are designed to reflect the amount of moisture available at a particular place and time, based on the amount of precipitation received as well as the temperature, which influences evaporation rates. The formula for creating this index was originally proposed in the 1960s (Palmer, 1965). Since then, the methods have been tested extensively and used to support hundreds of published studies. The PDSI is the most widespread and scientifically vetted drought index in use today.

The PDSI was designed to characterize long-term drought (i.e., patterns lasting a month or more). Because drought is cumulative, the formula takes precipitation and temperature data from previous weeks and months into account. Thus, a single rainy day is unlikely to cause a dramatic shift in the index.

PDSI values are normalized relative to long-term average conditions at each location, which means this method can be applied to any location regardless of how wet or dry it typically is. NOAA currently uses 1931–1990 as its long-term baseline. The index essentially measures deviation from normal conditions. The PDSI takes the form of a numerical value, generally ranging from -6 to +6. A value of zero reflects average conditions. Negative values indicate drier-than-average conditions and positive values indicate wetter-than-average conditions. NOAA provides the following interpretations for specific ranges of the index:

- 0 to -0.5 = normal
- -0.5 to -1.0 = incipient drought
- -1.0 to -2.0 = mild drought
- -2.0 to -3.0 = moderate drought
- -3.0 to -4.0 = severe drought
- < -4.0 = extreme drought

Similar adjectives can be applied to positive (wet) values.

NOAA calculates monthly values of the PDSI for each of the 344 climate divisions within the contiguous 48 states and corrects these data for time biases. These values are calculated from weather stations reporting both temperature and precipitation. All stations within a division are given equal weight. NOAA also combines PDSI values from all climate divisions to derive state-level averages for every month.

EPA obtained monthly state-level PDSI values from NOAA, then calculated annual averages for each state. To derive totals for the entire six-state region, EPA averaged the state-level data together for each year, weighted by state area. This procedure used state areas as defined by the U.S. Census Bureau at: https://www.census.gov/geo/reference/state-area.html.

To smooth out some of the year-to-year variability, EPA applied a nine-point binomial filter, which is plotted at the center of each nine-year window. For example, the smoothed value from 2005 to 2013 is plotted at year 2009. NOAA's NCDC recommends this approach. Figure 3 shows both the annual values and the smoothed curve.

EPA used endpoint padding to extend the nine-year smoothed lines all the way to the ends of the period of record. As recommended by NCDC, EPA calculated smoothed values as follows: if 2013 was the most recent year with data available, EPA calculated smoothed values to be centered at 2010, 2011, 2012, and 2013 by inserting the 2013 data point into the equation in place of the as-yet-unreported annual data points for 2014 and beyond. EPA used an equivalent approach at the beginning of the time series.

For more information about NOAA's processing methods, see the metadata file at: www7.ncdc.noaa.gov/CDO/DIV_DESC.txt. NOAA's website provides a variety of other references regarding the PDSI at: www.ncdc.noaa.gov/oa/climate/research/prelim/drought/palmer.html.

Feature Development

Various organizations define the Southwest in different ways—sometimes along political boundaries, sometimes along biogeographic or climatological boundaries. For this regional feature, EPA chose to focus on six states that are commonly thought of as "southwestern" and characterized at least in part by arid landscapes and scarce water supplies: Arizona, California, Colorado, Nevada, New Mexico, and Utah. EPA elected to follow state boundaries because several of the data sets are provided in the form of state averages, and because state boundaries are easily understood and relatable to a broad audience.

7. Quality Assurance and Quality Control

NOAA follows extensive quality assurance and quality control (QA/QC) procedures for collecting and compiling COOP weather station data. For documentation of COOP methods, including training manuals and maintenance of equipment, see: www.nws.noaa.gov/os/coop/training.htm. These training materials also discuss QC of the underlying data set. Pre-1948 COOP data were recently digitized from hard copy. Kunkel et al. (2005) discuss QC steps associated with digitization and other factors that might introduce error into an analysis.

When compiling NWS/COOP records into the *n*ClimDiv data set, NOAA employed a series of corrections to reduce potential biases. Steps include:

- Removal of duplicate records.
- Procedures to deal with missing data.
- Adjusting for changes in observing practices, such as changes in observation time.
- Testing and correcting for artificial discontinuities in a local station record, which might reflect station relocation, instrumentation changes, or urbanization (e.g., heat island effects).

For more information about these bias adjustments, see: www.ncdc.noaa.gov/monitoring-references/maps/us-climate-divisions.php and the references cited therein.

As described in NOAA's metadata file (www7.ncdc.noaa.gov/CDO/DIV_DESC.txt), the Time Bias Corrected Divisional PDSI data set has been adjusted to account for possible biases caused by differences in the time of reporting. A model by Karl et al. (1986) is used to adjust values so that all stations end their climatological day at midnight.

QA/QC procedures for Drought Monitor data are not readily available. Each underlying data source has its own methodology, which typically includes some degree of QA/QC. For example, precipitation and temperature data are verified and corrected by NOAA. Some of the other underlying data sources have QA/QC procedures available online, but others do not.

Analysis

8. Comparability Over Time and Space

Figures 1 and 3. Average Temperatures and Drought Severity

PDSI and temperature calculation methods, as obtained from NOAA's NWS/COOP data set, have been applied consistently over time and space. Although the equipment used may have varied, temperature readings are comparable for the entirety of the data set. The PDSI relies on the same underlying measurements (precipitation and temperature) in all cases. Although fewer stations were collecting weather data during the first few decades of the analysis, NOAA has determined that enough stations were available starting in 1895 to calculate valid index and temperature values for the six states presented in this regional feature.

Figure 2. Southwestern U.S. Lands Under Drought Conditions, 2000–2013

The resolution of the U.S. Drought Monitor has improved over time. When Drought Monitor calculations began, many of the component indicators used to determine drought conditions were reported at the climate division level. Many of these component indicators now include data from the county and sub-county level. This change in resolution over time can be seen in the methods used to draw contour lines on Drought Monitor maps.

The drought classification scheme used for the Drought Monitor is produced by combining data from several different sources. Different locations may use different primary sources, or they may use the same sources, but weighted differently. These data are combined to reflect the collective judgment of experts, and in some cases are adjusted to reconcile conflicting trends shown by different data sources over different time periods.

Though data resolution and mapping procedures have varied somewhat over time and space, the fundamental construction of the Drought Monitor has remained consistent.

9. Data Limitations

Factors that may impact the confidence, application, or conclusions drawn from this regional feature are as follows:

1. The feature gives a broad overview of drought conditions in the Southwest and is not intended to replace local or state information that might describe conditions more precisely. Local entities might monitor different variables to meet specific needs or to address local problems. As a consequence, there could be water shortages or crop failures within an area not designated as a drought area, just as there could be locations with adequate water supplies in an area designated as D3 or D4 (extreme or exceptional) drought.

2. Although the PDSI is arguably the most widely used drought index, it has some limitations that have been documented extensively in the literature. While the use of just two variables (precipitation and temperature) makes this index relatively easy to calculate over time and space, drought can have many other dimensions that these two variables do not fully capture. For example, the PDSI loses accuracy in areas where a substantial portion of the water supply comes from snowpack, which includes major portions of the Southwest.

3. Because this feature focuses on regional trends, it does not show how drought conditions vary by state or sub-state jurisdiction. For example, even if half of the Southwest suffered from severe drought, Figure 3 could show an average index value close to zero if the rest of the region was wetter than average. Thus, Figure 3 might understate the degree to which droughts are becoming more severe in some areas while other places receive more rain as a result of climate change.

4. Indices such as the U.S. Drought Monitor seek to address the limitations of the PDSI by incorporating many more variables. However, the Drought Monitor is relatively new and cannot yet be used to assess long-term climate trends. With several decades of data collection, future versions of Figure 2 should be able to paint a more complete picture of trends over time.

5. The drought classification scheme used for Figure 2 is produced by combining data from several different sources. These data are combined to reflect the collective judgment of experts and in some cases are adjusted to reconcile conflicting trends shown by different data sources over different time periods.

6. Uncertainties in surface temperature data increase as one goes back in time, as there are fewer stations earlier in the record. However, these uncertainties are not likely to mislead the user about fundamental trends in the data.

7. Biases in temperature measurements may have occurred as a result of changes over time in instrumentation, measuring procedures (e.g., time of day), and the exposure and location of the instruments. Where possible, data have been adjusted to account for changes in these variables. For more information on these corrections, see Section 7.

10. Sources of Uncertainty

Time biases for COOP temperature data are known to be small (< 0.3°F), while error estimates for the PDSI and Drought Monitor are unavailable. It is not clear how much uncertainty might be associated with the component indices that go into formulating the Drought Monitor or the process of compiling these indices into a single set of weekly values through averaging, weighting, and expert judgment.

11. Sources of Variability

Conditions associated with drought naturally vary from place to place and from one day to the next, depending on weather patterns and other factors. Figure 1 deliberately shows spatial variations, while addressing temporal variations through the use of multi-year averages. Figures 2 and 3 address spatial variability by presenting aggregate regional trends. Figure 2 smoothes out some of the inherent variability in drought measurement by relying on many indices, including several with a long-term focus. While Figure 2 shows noticeable week-to-week variability, it also reveals larger year-to-year patterns. Figure 3 addresses temporal variability by using an index that is designed to measure long-term drought and is not easily swayed by short-term conditions. Figure 3 also provides an annual average, along with a nine-year smoothed average.

12. Statistical/Trend Analysis

The statistical significance of the division-level temperature changes in Figure 1 has been assessed using ordinary least-squares linear regression of the annual data over the full period of record (1895–2013). Of the 38 climate divisions shown, all have positive long-term trends (i.e., increasing temperatures) that are significant at the 95 percent level.

Because data from the U.S. Drought Monitor (Figure 2) are only available for the most recent decade, this metric is too short-lived to be used for assessing long-term climate trends.

Ordinary least squares linear regression was used to estimate trends in drought according to the PDSI (Figure 3). For this six-state region as a whole, the long-term (1895–2013) trend is statistically significant at the 95 percent level (slope = -0.011 PDSI units per year; $p = 0.01$). Among individual states, Arizona and Nevada have experienced statistically significant trends toward more drought ($p < 0.05$). State-level results are shown in Table TD-1.

Table TD-1. State-Level Linear Regressions for PDSI Drought, 1895–2013

State	Slope	P-value
Arizona	-0.014	0.025
California	-0.009	0.075
Colorado	-0.010	0.123
Nevada	-0.013	0.045
New Mexico	-0.012	0.071
Utah	-0.012	0.062

References

Karl, T.R., C.N. Williams, Jr., P.J. Young, and W.M. Wendland. 1986. A model to estimate the time of observation bias associated with monthly mean maximum, minimum, and mean temperatures for the United States. J. Clim. Appl. Meteorol. 25(1):145–160.

Kunkel, K.E., D.R. Easterling, K. Hubbard, K. Redmond, K. Andsager, M.C. Kruk, and M.L. Spinar. 2005. Quality control of pre-1948 Cooperative Observer Network data. J. Atmos. Ocean. Tech. 22:1691–1705.

Palmer, W.C. 1965. Meteorological drought. Res. Paper No.45. Washington, DC: U.S. Department of Commerce.

Svoboda, M., D. Lecomte, M. Hayes, R. Heim, K. Gleason, J. Angel, B. Rippey, R. Tinker, M. Palecki, D. Stooksbury, D. Miskus, and S. Stephens. 2002. The drought monitor. B. Am. Meteorol. Soc. 83(8):1181–1190.

Willmott, C.J., and S.M. Robeson. 1995. Climatologically aided interpolation (CAI) of terrestrial air temperature. Int. J. Climatol. 15(2):221–229.

Tropical Cyclone Activity

Identification

1. Indicator Description

This indicator examines the aggregate activity of hurricanes and other tropical storms in the Atlantic Ocean, Caribbean, and Gulf of Mexico between 1878 and 2013. Climate change is expected to affect tropical cyclone activity through increased sea surface temperatures and other environmental changes that are key influences on cyclone formation and behavior.

Components of this indicator include:

- The number of hurricanes in the North Atlantic each year, along with the number making landfall in the United States (Figure 1).

- Frequency, intensity, and duration of North Atlantic cyclones as measured by the Accumulated Cyclone Energy Index (Figure 2).

- Frequency, intensity, and duration of North Atlantic cyclones as measured by the Power Dissipation Index (Figure 3).

2. Revision History

April 2010:	Indicator posted.
December 2011:	Updated Figure 2 with data through 2011.
April 2012:	Added hurricane counts (new Figure 1).
May 2012:	Updated Figure 3 with data through 2011.
March 2014:	Updated Figure 1 with data through 2013.
May 2014:	Updated Figures 2 and 3 with data through 2013.

Data Sources

3. Data Sources

This indicator is based on data maintained by the National Oceanic and Atmospheric Administration's (NOAA's) National Hurricane Center in a database referred to as HURDAT (HURricane DATa). This indicator presents three separate analyses of HURDAT data: a set of hurricane counts compiled by NOAA, NOAA's Accumulated Cyclone Energy (ACE) Index, and the Power Dissipation Index (PDI) developed by Dr. Kerry Emanuel at the Massachusetts Institute of Technology (MIT).

4. Data Availability

Figure 1. Number of Hurricanes in the North Atlantic, 1878–2013

Data for Figure 1 were obtained from several data sets published by NOAA:

- Total counts are available from NOAA's Atlantic Oceanographic and Meteorological Laboratory (AOML), Hurricane Research Division (HRD), at:
www.aoml.noaa.gov/hrd/hurdat/comparison_table.html.
- Landfalling counts are available from: www.aoml.noaa.gov/hrd/hurdat/comparison_table.html, with confirmation from www.aoml.noaa.gov/hrd/hurdat/All_U.S._Hurricanes.html.
- Adjusted counts for years prior to 1966 are based on a historical reanalysis posted by NOAA at:
www.gfdl.noaa.gov/index/cms-filesystem-action/user_files/gav/historical_storms/vk_11_hurricane_counts.txt (linked from:
www.gfdl.noaa.gov/gabriel-vecchi-noaa-gfdl).

Figure 2. North Atlantic Cyclone Intensity According to the Accumulated Cyclone Energy Index, 1950–2013

An overview of the ACE Index is available at:
www.cpc.ncep.noaa.gov/products/outlooks/background_information.shtml. The raw data for this indicator are published on NOAA's Hurricane Research Division website:
www.aoml.noaa.gov/hrd/hurdat/comparison_table.html.

Figure 3. North Atlantic Cyclone Intensity According to the Power Dissipation Index, 1949–2013

Emanuel (2005, 2007) gives an overview of the PDI, along with figures and tables. This indicator reports on an updated version of the data set (through 2013) that was provided by Dr. Kerry Emanuel.

Underlying Data

Wind speed measurements and other HURDAT data are available in various formats on NOAA's AOML website: www.aoml.noaa.gov/hrd/hurdat/Data_Storm.html. Since April 2014, NOAA has revised the format of the HURDAT data output, which is now called HURDAT2. Some documentation is available at: www.aoml.noaa.gov/hrd/hurdat/metadata_master.html, and definitions for the HURDAT2 data format are available at: www.aoml.noaa.gov/hrd/hurdat/newhurdat-format.pdf.

Methodology

5. Data Collection

This indicator is based on measurements of tropical cyclones over time. HURDAT compiles information on all hurricanes and other tropical storms occurring in the North Atlantic Ocean, including parameters such as wind speed, barometric pressure, storm tracks, and dates. Field methods for data collection and analysis are documented in official NOAA publications (Jarvinen et al., 1984). This indicator is based on sustained wind speed, which is defined as the one-minute average wind speed at an altitude of 10 meters.

Data collection methods have evolved over time. When data collection began, ships and land observation stations were used to measure and track storms. Analysts compiled all available wind speed observations and all information about the measurement technique to determine the wind speed for the four daily intervals for which the storm track was recorded.

More recently, organized aircraft reconnaissance, the coastal radar network, and weather satellites with visible and infrared sensors have improved accuracy in determining storm track, maximum wind speeds, and other storm parameters, such as central pressure. Weather satellites were first used in the 1960s to detect the initial position of a storm system; reconnaissance aircraft would then fly to the location to collect precise measurements of the wind field, central pressure, and location of the center. Data collection methods have since improved with more sophisticated satellites.

This indicator covers storms occurring in the Atlantic Ocean north of the equator, including the Caribbean Sea and the Gulf of Mexico. In addition to tropical storms, HURDAT2 includes data from storms classified as extratropical and subtropical, although extratropical storms are not counted in this indicator. Subtropical cyclones exhibit some characteristics of a tropical cyclone but also some characteristics of an extratropical storm. Subtropical cyclones are now named in conjunction with the tropical storm naming scheme, and in practice, many subtropical storms eventually turn into tropical storms. HURDAT2 is updated annually by NOAA and data are available from 1878 through 2013.

Sampling and analysis procedures for the HURDAT data are described by Jarvinen et al. (1984) for collection methods up to 1984. Changes to past collection methods are partially described in the supplementary methods from Emanuel (2005). Other data explanations are available at: www.nhc.noaa.gov/pastall.shtml#hurdat. The mission catalogue of data sets collected by NOAA aircraft is available at: www.aoml.noaa.gov/hrd/data_sub/hurr.html.

6. Indicator Derivation

Figure 1. Number of Hurricanes in the North Atlantic, 1878–2013

This figure displays three time series: the number of hurricanes per year making landfall in the United States, the total number of hurricanes on record for the North Atlantic, and an adjusted total that attempts to account for changes in observing capabilities. All three counts are limited to cyclones in the North Atlantic (i.e., north of the equator) meeting the definition of a hurricane, which requires sustained wind speeds of at least 74 miles per hour.

Landfalling counts reflect the following considerations:

- If a single hurricane made multiple U.S. landfalls, it is only counted once.
- If the hurricane center did not make a U.S. landfall (or substantially weakened before making landfall), but did produce hurricane-force winds over land, it is counted.
- If the hurricane center made landfall in Mexico, but did produce hurricane-force winds over the United States, it is counted.
- If a storm center made a U.S. landfall, but all hurricane-force winds (if any) remained offshore, it is not counted. This criterion excludes one storm in 1888 and another in 1908.

For all years prior to the onset of complete satellite coverage in 1966, total basin-wide counts have been adjusted upward based on historical records of ship track density. In other words, during years when fewer ships were making observations in a given ocean region, hurricanes in that region were more likely to have been missed, or their intensity underestimated to be below hurricane strength, leading to a larger corresponding adjustment to the count for those years. These adjustment methods are cited in Knutson et al. (2010) and described in more detail by Vecchi and Knutson (2008), Landsea et al. (2010), and Vecchi and Knutson (2011).

The overall adjustment process can be described by the simple formula $x + y = z$, where:

- x = raw total (number of hurricanes) from HURDAT
- y = adjustment factor
- z = adjusted total

NOAA provided adjusted totals (z) in 2012, which EPA converted to adjustment factors (y) by subtracting the corresponding raw totals that were available from HURDAT at the time (x). This step was needed because historical raw totals are subject to change as HURDAT is reanalyzed and improved over time. For example, between summer 2012 and spring 2013, raw hurricane counts changed for 11 years in HURDAT. Most of these cases occurred prior to 1940, and almost all involved an increase or decrease of only one storm in a given year. The adjustment factors (y) do not need to change, as they were calculated by comparing post-1965 storms against ship tracks for pre-1966 years, and neither of these variables is changing as a result of ongoing HURDAT revisions. Thus, where HURDAT reanalysis resulted in a new (x), EPA added the previously determined (y), leading to a new (z). This approach was recommended by the NOAA data providers.

All three curves have been smoothed using a five-year unweighted average, as recommended by the data provider. Data are plotted at the center of each window; for example, the five-year smoothed value for 1949 to 1953 is plotted at year 1951. Because of this smoothing procedure and the absence of endpoint padding, no averages can be plotted for the first two years and last two years of the period of record (1878, 1879, 2012, and 2013).

Figure 2. North Atlantic Tropical Cyclone Activity According to the Accumulated Cyclone Energy Index, 1950–2013

This figure uses NOAA's ACE Index to describe the combined frequency, strength, and duration of tropical storms and hurricanes each season. As described by Bell and Chelliah (2006), "the ACE Index is calculated by summing the squares of the estimated 6-hourly maximum sustained wind speed in knots for all periods while the system is either a tropical storm or hurricane." A system is considered at least a tropical storm if it has a wind speed of at least 39 miles per hour. The ACE Index is preferred over other similar indices such as the Hurricane Destruction Potential (HDP) and the Net Tropical Cyclone Index (NTC) because it takes tropical storms into account and it does not include multiple sampling of some parameters. The ACE Index also includes subtropical cyclones, which are named using the same scheme as tropical cyclones and may eventually turn into tropical cyclones in some cases. The index does not include information on storm size, which is an important component of a storm's damage potential.

Figure 2 of the indicator shows annual values of the ACE, which are determined by summing the individual ACE Index values of all storms during that year. The index itself is measured in units of wind speed squared, but for this indicator, the index has been converted to a numerical scale where 100 equals the median value over a base period from 1981 to 2010. A value of 150 would therefore represent 150 percent of the median, or 50 percent more than normal. NOAA has also established a set of thresholds to categorize each hurricane season as "above normal," "near normal," or "below normal" based on the distribution of observed values during the base period. The "near normal" range extends from 71.4 to 120 percent of the median, with the "above normal" range above 120 percent of the median and the "below normal" range below 71.4 percent.

ACE Index computation methods and seasonal classifications are described by Bell and Chelliah (2006). This information is also available on the NOAA website at:
www.cpc.noaa.gov/products/outlooks/background_information.shtml.

Figure 3. North Atlantic Tropical Cyclone Activity According to the Power Dissipation Index, 1949–2013

For additional perspective, this figure presents the PDI. Like the ACE Index, the PDI is also based on storm frequency, wind speed, and duration, but it uses a different calculation method that places more emphasis on storm intensity by using the cube of the wind speed rather than the wind speed squared (as for the ACE). Emanuel (2005, 2007) provides a complete description of how the PDI is calculated. Emanuel (2007) also explains adjustments that were made to correct for biases in the quality of storm observations and wind speed measurements early in the period of record. The PDI data in Figure 3 of this indicator are in units of 10^{11} m^3/s^2, but the actual figure omits this unit and simply alludes to "index values" in order to make the indicator accessible to the broadest possible audience.

The PDI data shown in Figure 3 have been smoothed using a five-year weighted average applied with weights of 1, 3, 4, 3, and 1. This method applies greater weight to values near the center of each five-year window. Data are plotted at the center of each window; for example, the five-year smoothed value for 1949 to 1953 is plotted at year 1951. The data providers recommend against endpoint padding for these particular variables, based on past experience and their expert judgment, so no averages can be plotted for the first two years and last two years of the period of record (1949, 1950, 2012, and 2013).

The PDI includes all storms that are in the so-called "best track" data set issued by NOAA, which can include subtropical storms. Weak storms contribute very little to power dissipation, however, so subtropical storms typically have little impact on the final metric.

Emanuel (2005, 2007) describes methods for calculating the PDI and deriving the underlying power dissipation formulas. Analysis techniques, data sources, and corrections to raw data used to compute the PDI are described in the supplementary methods for Emanuel (2005), with further corrections addressed in Emanuel (2007).

Sea surface temperature has been plotted for reference, based on methods described in Emanuel (2005, 2007). The curve in Figure 3 represents average sea surface temperature in the area of storm genesis in the North Atlantic: specifically, a box bounded in latitude by 6°N and 18°N, and in longitude by 20°W and 60°W. Values have been smoothed over five-year periods. For the sake of straightforward presentation, sea surface temperature has been plotted in unitless form without a secondary axis, and the curve has been positioned to clearly show the relationship between sea surface temperature and the PDI.

7. Quality Assurance and Quality Control

Jarvinen et al. (1984) describe quality assurance/quality control procedures for each of the variables in the HURDAT data set. Corrections to early HURDAT data are made on an ongoing basis through the HURDAT re-analysis project to correct for both systematic and random errors identified in the data set. Information on this re-analysis is available at on the NOAA website at: www.aoml.noaa.gov/hrd/data_sub/re_anal.html. Emanuel (2005) provides a supplementary methods document that describes both the evolution of more accurate sample collection technology and further corrections made to the data.

Analysis

8. Comparability Over Time and Space

In the early years of the data set there is a high likelihood that some tropical storms went undetected, as observations of storms were made only by ships at sea and land-based stations. Storm detection improved over time as ship track density increased, and, beginning in 1944, with the use of organized aircraft reconnaissance (Jarvinen et al., 1984). However, it was not until the late 1960s and later, when satellite coverage was generally available, that the Atlantic tropical cyclone frequency record can be assumed to be relatively complete. Because of the greater uncertainties inherent in earlier data, Figure 1 adjusts pre-1966 data to account for the density of ship observations, while Figures 2 and 3 exclude data prior to 1950 and 1949, respectively. If the best available science warrants, NOAA occasionally re-analyzes historical HURDAT data (www.aoml.noaa.gov/hrd/data_sub/re_anal.html) to adjust for both random and systematic error present in data from the beginning of the time series. Most of these changes affect data prior to 1950, but NOAA has also revised more recent ACE Index values slightly.

Emanuel (2005) describes the evolution of more accurate sample collection technology and various corrections made to the data. For the PDI, Emanuel (2007) employed an additional bias correction process for the early part of the period of record (the 1950s and 1960s), when aircraft reconnaissance and radar technology were less robust than they are today—possibly resulting in missed storms or underestimated power. These additional corrections were prompted in part by an analysis published by Landsea (1993).

9. Data Limitations

Factors that may impact the confidence, application, or conclusions drawn from this indicator are as follows:

1. Methods of detecting hurricanes have improved over time, and raw counts prior to the 1960s may undercount the total number of hurricanes that formed each year. However, Figure 1 presents an adjusted time series to attempt to address this limitation.

2. Wind speeds are measured using several observation methods with varying levels of uncertainty, and these methods have improved over time. The wind speeds recorded in HURDAT should be considered the best estimate of several wind speed observations compiled by analysts.

3. Many different indices have been developed to analyze storm duration, intensity, and threat. Each index has strengths and weaknesses associated with its ability to describe these parameters. The indices used in this indicator (hurricane counts, ACE Index, and PDI) are considered to be among the most reliable.

10. Sources of Uncertainty

Counts of landfalling U.S. hurricanes are considered reliable back to the late 1800s, as population centers and recordkeeping were present all along the Gulf and Atlantic coasts at the time. Total hurricane counts for the North Atlantic became fairly reliable after aircraft reconnaissance began in 1944, and became highly reliable after the onset of satellite tracking around 1966. Prior to the use of these two methods, however, detection of non-landfalling storms depended on observations from ships, which could lead to undercounting due to low density of ship coverage. Figure 1 shows how pre-1966 counts have been adjusted upward based on the density of ship tracks (Vecchi and Knutson, 2011).

The ACE Index and the PDI are calculated directly from wind speed measurements. Thus, the main source of possible uncertainty in the indicator is within the underlying HURDAT data set. Because the determination of storm track and wind speed requires some expert judgment by analysts, some uncertainty is likely. Methodological improvements suggest that recent data may be somewhat more accurate than earlier measurements. Landsea and Franklin (2013) have estimated the average uncertainty for measurements of intensity, central pressure, position, and size of Atlantic hurricanes in recent years. They also compare present-day uncertainty with uncertainty estimates from the 1990s. Uncertainty estimates for older HURDAT data are not readily available.

Because uncertainty varies depending on observation method, and these methods have evolved over time, it is difficult to make a definitive statement about the impact of uncertainty on Figures 2 and 3. Changes in data gathering technologies could substantially influence the overall patterns in Figures 2 and 3, and the effects of these changes on data consistency over the life of the indicator would benefit from additional research.

11. Sources of Variability

Intensity varies by storm and location. The indicator addresses this type of variability by using two indices that aggregate all North Atlantic storms within a given year. Aggregate annual intensity also varies from year to year as a result of normal variation in weather patterns, multi-year climate cycles, and other factors. Annual storm counts can vary from year to year for similar reasons. Figure 2 shows interannual variability. Figures 1 and 3 also show variability over time, but they seek to focus on longer-term variability and trends by presenting a five-year smoothed curve.

Overall, it remains uncertain whether past changes in any tropical cyclone activity (frequency, intensity, rainfall, and so on) exceed the variability expected through natural causes, after accounting for changes over time in observing capabilities (Knutson et al., 2010).

12. Statistical/Trend Analysis

This indicator does not report on the slope of the apparent trends in hurricane counts or cyclone intensity, nor does it calculate the statistical significance of these trends. See Vecchi and Knutson (2008, 2011) for examples of such a trend analysis, including statistical significance tests.

References

Bell, G.D., and M. Chelliah. 2006. Leading tropical modes associated with interannual and multidecadal fluctuations in North Atlantic hurricane activity. J. Climate 19:590–612.

Emanuel, K. 2005. Increasing destructiveness of tropical cyclones over the past 30 years. Nature 436:686–688. Supplementary methods available with the online version of the paper at: www.nature.com/nature/journal/v436/n7051/suppinfo/nature03906.html.

Emanuel, K. 2007. Environmental factors affecting tropical cyclone power dissipation. J. Climate 20(22):5497–5509. ftp://texmex.mit.edu/pub/emanuel/PAPERS/Factors.pdf.

Jarvinen, B.R., C.J. Neumann, and M.A.S. Davis. 1984. A tropical cyclone data tape for the North Atlantic Basin, 1886–1983: Contents, limitations and uses. NOAA Technical Memo NWS NHC 22.

Knutson, T.R., J.L. McBride, J. Chan, K. Emanuel, G. Holland, C. Landsea, I. Held, J.P. Kossin, A.K. Srivastava, and M. Sugi. 2010. Tropical cyclones and climate change. Nat. Geosci. 3: 157–163.

Landsea, C.W. 1993. A climatology of intense (or major) Atlantic hurricanes. Mon. Weather Rev. 121:1703–1713. www.aoml.noaa.gov/hrd/Landsea/Landsea_MWRJune1993.pdf.

Landsea, C., G.A. Vecchi, L. Bengtsson, and T.R. Knutson. 2010. Impact of duration thresholds on Atlantic tropical cyclone counts. J. Climate 23:2508–2519. www.nhc.noaa.gov/pdf/landsea-et-al-jclim2010.pdf.

Landsea, C., and J. Franklin. 2013. Atlantic hurricane database uncertainty and presentation of a new database format. Mon. Weather Rev. 141:3576–3592.

Vecchi, G.A., and T.R. Knutson. 2008. On estimates of historical North Atlantic tropical cyclone activity. J. Climate 21:3580–3600.

Vecchi, G.A., and T.R. Knutson. 2011. Estimating annual numbers of Atlantic hurricanes missing from the HURDAT database (1878–1965) using ship track density. J. Climate 24(6):1736–1746. www.gfdl.noaa.gov/bibliography/related_files/gav_2010JCLI3810.pdf.

Ocean Heat

Identification

1. Indicator Description

This indicator describes trends in the amount of heat stored in the world's oceans between 1955 and 2013. The amount of heat in the ocean, or ocean heat content, is an important indicator of climate change because the oceans ultimately absorb a large portion of the extra energy that greenhouse gases trap near the Earth's surface. Ocean heat content also plays an important role in the Earth's climate system because heat from ocean surface waters provides energy for storms and thereby influences weather patterns.

2. Revision History

April 2010:   Indicator posted.
April 2012:   Updated with data through 2011.
August 2013:   Updated on EPA's website with data through 2012.
April 2014:   Updated with data through 2013.

Data Sources

3. Data Sources

This indicator is based on analyses conducted by three different government agencies:
- Australia's Commonwealth Scientific and Industrial Research Organisation (CSIRO)
- Japan Meteorological Agency's Meteorological Research Institute (MRI/JMA)
- National Oceanic and Atmospheric Administration (NOAA)

MRI/JMA used four different data sets: the World Ocean Database (WOD), the World Ocean Atlas (WOA), the Global Temperature-Salinity Profile Program (GTSPP, which was used to fill gaps in the WOD since 1990), and data from the Japan Maritime Self-Defense Force (JMSDF). CSIRO used two data sets: ocean temperature profiles in the ENACT/ENSEMBLES version 3 (EN3) and data collected using 60,000 Argo profiling floats. Additionally, CSIRO included bias-corrected Argo data, as described in Barker et al. (2011), and bias-corrected expendable bathythermograph (XBT) data from Wijffels et al. (2008). NOAA also used data from the WOD and WOA.

4. Data Availability

EPA developed Figure 1 using trend data from three ongoing studies. Data and documentation from these studies can be found at the following links:

- CSIRO: www.cmar.csiro.au/sealevel/thermal_expansion_ocean_heat_timeseries.html. Select "GOHC_recons_version3.1_1950_2012_CLIM_sbca12tmosme_OBS_bcax_0700m.dat" to download the data. See additional documentation in Domingues et al. (2008).

- MRI/JMA: Data from Ishii and Kimoto (2009) are posted at: http://atm-phys.nies.go.jp/~ism/pub/ProjD/doc. Updated data were provided by the author, Masayoshi Ishii. Data are expected to be updated regularly online in the future. See additional documentation in Ishii and Kimoto (2009).

- NOAA: www.nodc.noaa.gov/OC5/3M_HEAT_CONTENT. Under "Heat Content, " select "basin time series." Then, under "Yearly heat content from 1955 to 2013," select the "0– 700" file under "World." See additional documentation in Levitus et al. (2009).

The underlying data for this indicator come from a variety of sources. Some of these data sets are publicly available, but other data sets consist of samples gathered by the authors of the source papers, and these data might be more difficult to obtain online. WOA and WOD data and descriptions of data are available on NOAA's National Oceanographic Data Center (NODC) website at: www.nodc.noaa.gov.

Methodology

5. Data Collection

This indicator reports on the amount of heat stored in the ocean from sea level to a depth of 700 meters, which accounts for approximately 17.5 percent of the total global ocean volume (calculation from Catia Domingues, CSIRO). Each of the three studies used to develop this indicator uses several ocean temperature profile data sets to calculate an ocean heat content trend line.

Several different devices are used to sample temperature profiles in the ocean. Primary methods used to collect data for this indicator include XBT; mechanical bathythermographs (MBT); Argo profiling floats; reversing thermometers; and conductivity, temperature, and depth sensors (CTD). These instruments produce temperature profile measurements of the ocean water column by recording data on temperature and depth. The exact methods used to record temperature and depth vary. For instance, XBTs use a fall rate equation to determine depth, whereas other devices measure depth directly.

Each of the three studies used to develop this indicator relies on different combinations of devices; for example, the CSIRO analysis excludes MBT data. More information on the three main studies and their respective methods can be found at:

- CSIRO: Domingues et al. (2008) and: www.cmar.csiro.au/sealevel/thermal_expansion_ocean_heat_timeseries.html.

- MRI/JMA: Ishii and Kimoto (2009) and: http://atm-phys.nies.go.jp/~ism/pub/ProjD/doc.

- NOAA: Levitus et al. (2009) and: www.nodc.noaa.gov/OC5/3M_HEAT_CONTENT.

Studies that measure ocean temperature profiles are generally designed using in situ oceanographic observations and analyzed over a defined and spatially uniform grid (Ishii and Kimoto, 2009). For instance, the WOA data set consists of in situ measurements of climatological fields, including temperature, measured in a 1-degree grid. Sampling procedures for WOD and WOA data are provided by NOAA's NODC at: www.nodc.noaa.gov/OC5/indprod.html. More information on the WOA sample design in particular can be found at: www.nodc.noaa.gov/OC5/WOA05/pr_woa05.html.

At the time of last update, data from all three sources were available through 2013.

6. Indicator Derivation

While details of data analysis are particular to the individual study, in general, temperature profile data were averaged monthly at specific depths within rectangular grid cells. In some cases, interpolation techniques were used to fill gaps where observational spatial coverage was sparse. Additional steps were taken to correct for known biases in XBT data. Finally, temperature observations were used to calculate ocean heat content through various conversions. The model used to convert measurements was consistent across all three studies cited by this indicator.

Barker et al. (2011) describe instrument biases and procedures for correcting for these biases. For more information about interpolation and other analytical steps, see Ishii and Kimoto (2009), Domingues et al. (2008), Levitus et al. (2009), and references therein.

Each study used a different long-term average as a baseline. To allow more consistent comparison, EPA adjusted each curve such that its 1971–2000 average would be set at zero. Choosing a different baseline period would not change the shape of the data over time. Although some of the studies had pre-1955 data, Figure 1 begins at 1955 for consistency. The current CSIRO data series is based on updates to the original data set provided in Domingues et al. (2008) and plotted with a start date of 1960. The updated data set excludes 1955–1959, as the authors (Domingues et al.) have expressed diminished confidence in their data set for this period because there are fewer ocean observations in the early part of the record.

7. Quality Assurance and Quality Control

Data collection and archival steps included QA/QC procedures. For example, QA/QC measures for the WOA are available at: ftp://ftp.nodc.noaa.gov/pub/data.nodc/woa/PUBLICATIONS/qc94tso.pdf. Each of the data collection techniques involves different QA/QC measures. For example, a summary of studies concerning QA/QC of XBT data is available from NODC at: www.nodc.noaa.gov/OC5/XBT_BIAS/xbt_bibliography.html. The same site also provides additional information about QA/QC of ocean heat data made available by NODC.

All of the analyses performed for this indicator included additional QA/QC steps at the analytical stage. In each of the three main studies used in this indicator, the authors carefully describe QA/QC methods, or provide the relevant references.

Analysis

8. Comparability Over Time and Space

Analysis of raw data is complicated because data come from a variety of observational methods, and each observational method requires certain corrections to be made. For example, systematic biases in XBT depth measurements have recently been identified. These biases were shown to lead to erroneous estimates of ocean heat content through time. Each of the three main studies used in this indicator corrects for these XBT biases. Correction methods are slightly different among studies and are described in detail in each respective paper. More information on newly identified biases associated with XBT can be found in Barker et al. (2011).

This indicator presents three separate trend lines to compare different estimates of ocean heat content over time. Each estimate is based on analytical methods that have been applied consistently over time and space. General agreement among trend lines, despite some year-to-year variability, indicates a robust trend.

9. Data Limitations

Factors that may impact the confidence, application, or conclusions drawn from this indicator are as follows:

1. Data must be carefully reconstructed and filtered for biases because of different data collection techniques and uneven sampling over time and space. Various methods of correcting the data have led to slightly different versions of the ocean heat trend line.

2. In addition to differences among methods, some biases may be inherent in certain methods. The older MBT and XBT technologies have the highest uncertainty associated with measurements.

3. Limitations of data collection over time and especially over space affect the accuracy of observations. In some cases, interpolation procedures were used to complete data sets that were spatially sparse.

10. Sources of Uncertainty

Uncertainty measurements can be made by the organizations responsible for data collection, and they can also be made during subsequent analysis. One example of uncertainty measurements performed by an agency is available for the WOA at: www.nodc.noaa.gov/OC5/indprod.html.

Error estimates associated with each of the curves in Figure 1 are discussed in Domingues et al. (2008), Ishii and Kimoto (2009), and Levitus et al. (2009). All of the data files listed in Section 4 ("Data Availability") include a one-sigma error value for each year.

11. Sources of Variability

Weather patterns, seasonal changes, multiyear climate oscillations, and many other factors could lead to day-to-day and year-to-year variability in ocean temperature measurements at a given location. This indicator addresses some of these forms of variability by aggregating data over time and space to calculate annual values for global ocean heat content. The overall increase in ocean heat over time (as shown by all three analyses) far exceeds the range of interannual variability in ocean heat estimates.

12. Statistical/Trend Analysis

Domingues et al. (2008), Ishii and Kimoto (2009), and Levitus et al. (2009) have all calculated linear trends and corresponding error values for their respective ocean heat time series. Exact time frames and slopes vary among the three publications, but they all reveal a generally upward trend (i.e., increasing ocean heat over time).

References

Barker, P.M., J.R. Dunn, C.M. Domingues, and S.E. Wijffels. 2011. Pressure sensor drifts in Argo and their impacts. J. Atmos. Oceanic Tech. 28:1036–1049.

Domingues, C.M., J.A. Church, N.J. White, P.J. Gleckler, S.E. Wijffels, P.M. Barker, and J.R. Dunn. 2008. Improved estimates of upper-ocean warming and multi-decadal sea-level rise. Nature 453:1090–1093.

Ishii, M., and M. Kimoto. 2009. Reevaluation of historical ocean heat content variations with time-varying XBT and MBT depth bias corrections. J. Oceanogr. 65:287–299.

Levitus, S., J.I. Antonov, T.P. Boyer, R.A. Locarnini, H.E. Garcia, and A.V. Mishonov. 2009. Global ocean heat content 1955–2008 in light of recently revealed instrumentation problems. Geophys. Res. Lett. 36:L07608.

Wijffels, S.E., J. Willis, C.M. Domingues, P. Barker, N.J. White, A. Gronell, K. Ridgway, and J.A. Church. 2008. Changing expendable bathythermograph fall rates and their impact on estimates of thermosteric sea level rise. J. Climate 21:5657–5672.

Sea Surface Temperature

Identification

1. Indicator Description

This indicator describes global trends in sea surface temperature (SST) from 1880 to 2013. SST is a key indicator related to climate change because it describes conditions at the boundary between the atmosphere and the oceans, which is where the transfer of energy between the atmosphere and oceans takes place. As the oceans absorb more heat from the atmosphere, SST is expected to increase. Changes in SST can affect circulation patterns and ecosystems in the ocean, and they can also influence global climate through the transfer of energy back to the atmosphere.

Components of this indicator include:

- Global average SST from 1880 to 2013 (Figure 1)
- A global map showing variations in SST change from 1901 to 2012 (Figure 2)

2. Revision History

April 2010:	Indicator posted.
December 2011:	Updated with data through 2010.
January 2012:	Updated with data through 2011.
April 2012:	Updated with revised data through 2011.
July 2012:	Updated example map.
August 2013:	Updated Figure 1 on EPA's website with data through 2012.
December 2013:	Replaced example map with new map of change over time (Figure 2).
March 2014:	Updated Figure 1 with revised data through 2013.

Data Sources

3. Data Sources

Figure 1 is based on the Extended Reconstructed Sea Surface Temperature (ERSST) analysis developed by the National Oceanic and Atmospheric Administration's (NOAA's) National Climatic Data Center (NCDC). The reconstruction model used here is ERSST version 3b (ERSST.v3b), which covers the years 1880 to 2013 and was described in Smith et al. (2008).

Figure 2 has been adapted from a map in the Intergovernmental Panel on Climate Change's (IPCC's) Fifth Assessment Report (IPCC, 2013). The original map appears in IPCC's Working Group I report as Figure SPM.1 and Figure 2.21, and it shows temperature change over land as well as over the ocean. The data originally come from NCDC's Merged Land-Ocean Surface Temperature Analysis (MLOST), which combines land-surface air temperature data with SST data from ERSST.v3b.

ERSST.v3b is based on a large set of temperature measurements dating back to the 1800s. This data set is called the International Comprehensive Ocean-Atmosphere Data Set (ICOADS), and it is compiled and maintained by NOAA.

4. Data Availability

NCDC and the National Center for Atmospheric Research (NCAR) provide access to monthly and annual SST and error data from the ERSST.v3b reconstruction in Figure 1, as well as a mapping utility that allows the user to calculate average anomalies over time and space (NOAA, 2014a). EPA used global data (all latitudes), which can be downloaded from: ftp://ftp.ncdc.noaa.gov/pub/data/mlost/operational. Specifically, EPA used the ASCII text file "aravg.ann.ocean.90S.90N.asc", which includes annual anomalies and error variance. A "readme" file in the same FTP directory explains how to use the ASCII file. The ERSST.v3b reconstruction is based on in situ measurements, which are available online through the ICOADS (NOAA, 2014b).

Figure 2 was taken directly from IPCC (2013). Underlying gridded data and documentation are available at: ftp://ftp.ncdc.noaa.gov/pub/data/mlost/operational.

Underlying ICOADS data are available at: http://icoads.noaa.gov.

Methodology

5. Data Collection

Both components of this indicator—global average SST since 1880 and the map of variations in SST change since 1901—are based on in situ instrumental measurements of water temperature worldwide. When paired with appropriate screening criteria and bias correction algorithms, in situ records provide a reliable long-term record of temperature. The long-term sampling was not based on a scientific sampling design, but was gathered by "ships of opportunity" and other ad hoc records. Records were particularly sparse or problematic before 1900 and during the two World Wars. Since about 1955, in situ sampling has become more systematic and measurement methods have continued to improve. SST observations from drifting and moored buoys were first used in the late 1970s. Buoy observations became more plentiful following the start of the Tropical Ocean Global Atmosphere (TOGA) program in 1985. Locations have been selected to fill in data gaps where ship observations are sparse.

A summary of the relative availability, coverage, accuracy, and biases of the different measurement methods is provided by Reynolds et al. (2002). Sampling and analytical procedures are documented in several publications that can be accessed online. NOAA has documented the measurement, compilation, quality assurance, editing, and analysis of the underlying ICOADS sea surface data set at: http://icoads.noaa.gov/publications.html.

Although SST can also be interpreted from satellite imagery, ERSST.v3b does not include satellite data. In the original update from ERSST v2 to v3, satellite data were added to the analysis. However, ERSST.v3b no longer includes satellite data because the addition of satellite data caused problems for many users. Although the satellite data were corrected with respect to the in situ data, a residual cold bias remained. The bias was strongest in the middle and high latitude Southern Hemisphere where in situ data were sparse. The residual bias led to a modest decrease in the global warming trend and modified global annual temperature rankings.

6. Indicator Derivation

Figure 1. Average Global Sea Surface Temperature, 1880–2013

This figure is based on the ERSST, a reconstruction of historical SST using in situ data. The reconstruction methodology has undergone several stages of development and refinement. This figure is based on the most recent data release, version 3b (ERSST.v3b).

This reconstruction involves filtering and blending data sets that use alternative measurement methods and include redundancies in space and time. Because of these redundancies, this reconstruction is able to fill spatial and temporal data gaps and correct for biases in the different measurement techniques (e.g., uninsulated canvas buckets, intakes near warm engines, uneven spatial coverage). Locations have been combined to report a single global value, based on scientifically valid techniques for averaging over areas. Specifically, data have been averaged over 5-by-5-degree grid cells as part of the MLOST data product (www.esrl.noaa.gov/psd/data/gridded/data.mlost.html). Daily and monthly records have been averaged to find annual anomalies. Thus, the combined set of measurements is stronger than any single set. Reconstruction methods are documented in more detail by Smith et al. (2008). Smith and Reynolds (2005) discuss and analyze the similarities and differences between various reconstructions, showing that the results are generally consistent. For example, the long-term average change obtained by this method is very similar to those of the "unanalyzed" measurements and reconstructions discussed by Rayner et al. (2003).

This figure shows the extended reconstructed data as anomalies, or differences, from a baseline "climate normal." In this case, the climate normal was defined to be the average SST from 1971 to 2000. No attempt was made to project data beyond the period during which measurements were collected.

Additional information on the compilation, data screening, reconstruction, and error analysis of the reconstructed SST data can be found at: http://www.ncdc.noaa.gov/ersst.

Figure 2. Change in Sea Surface Temperature, 1901–2012

This map is based on gridded data from MLOST, which in turn draws SST data from ERSST.v3. ERSST's analytical methods are described above for Figure 1.

EPA replicated the map in IPCC (2013) by calculating trends for each grid cell using the same Interactive Data Language (IDL) code that the authors of IPCC (2013) used. A long-term trend was calculated for each grid cell using linear regression. Trends have been calculated only for those cells with more than 70 percent complete records and more than 20 percent data availability during the first and last 10 percent of years (i.e., the first and last 11 years). The slope of each grid cell's trend (i.e., the rate of change per year) was multiplied by the number of years in the period to derive an estimate of total change. Parts of the ocean that are blank on the map did not meet these data availability thresholds. Black plus signs (+) indicate grid cells where the long-term trend is significant to a 90 percent confidence level.

EPA re-plotted IPCC's map and displayed only the ocean pixels (no land-based data) because this indicator focuses on SST. EPA also converted the results from Celsius to Fahrenheit.

7. Quality Assurance and Quality Control

Thorough documentation of the quality assurance and quality control (QA/QC) methods and results is available in the technical references for ERSST.v3b (www.ncdc.noaa.gov/ersst).

Analysis

8. Comparability Over Time and Space

Presenting the data at a global and annual scale reduces the uncertainty and variability inherent in SST measurements, and therefore the overall reconstruction in Figure 1 is considered to be a good representation of global SST. This data set covers the Earth's oceans with sufficient frequency and resolution to ensure that overall averages are not inappropriately distorted by singular events or missing data due to sparse in situ measurements or cloud cover. The confidence interval shows the degree of accuracy associated with the estimates over time and suggests that later measurements may be used with greater confidence than pre-20th century estimates.

Figure 2 is based on several data products that have been carefully assembled to maximize consistency over time and space. Areas with insufficient data for calculating trends have been excluded from the map.

Continuous improvement and greater spatial resolution can be expected in the coming years as historical data are updated. For example, there is a known bias during the World War II years (1941–1945), when almost all measurements were collected by U.S. Navy ships that recorded ocean intake temperatures, which can give warmer numbers than the techniques used in other years. Future efforts will adjust the data more suitably to account for this bias.

Researchers Smith and Reynolds (2005) have compared ERSST.v3b with other similar reconstructions using alternative methods. These comparisons yield consistent results, albeit with narrower uncertainty estimates. Hence, the graph presented in Figure 1 may be more conservative than would be the case had alternative methods been employed.

9. Data Limitations

Factors that may impact the confidence, application, or conclusions drawn from this indicator are as follows:

1. The 95 percent confidence interval in Figure 1 is wider than in other methods for long-term reconstructions; in mean SSTs, this interval tends to dampen anomalies.

2. The geographic resolution of Figure 1 is coarse for ecosystem analyses, but reflects long-term and global changes as well as shorter-term variability.

3. The reconstruction methods used to create both components of this indicator remove most random noise in the data. However, the anomalies are also dampened when and where data are too sparse for a reliable reconstruction. The 95 percent confidence interval in Figure 1 reflects this dampening effect and uncertainty caused by possible biases in the observations.

4. Data screening results in loss of multiple observations at latitudes higher than 60 degrees north or south. Effects of screening at high latitudes are minimal in the context of the global average; the main effect is to lessen anomalies and widen confidence intervals. This screening does create gaps in the Figure 2 map, however.

10. Sources of Uncertainty

The ERSST.v3b model has largely corrected for measurement error, but some uncertainty still exists. Contributing factors include variations in sampling methodology by era as well as geographic region, and instrument error from both buoys as well as ships.

The ERSST.v3b global reconstruction (Figure 1) includes an error variance for each year, which is associated with the biases and errors in the measurements and treatments of the data. NOAA has separated this variance into three components: high-frequency error, low-frequency error, and bias error. For this indicator, the total variance was used to calculate a 95-percent confidence interval (see Figure 1) so that the user can understand the impact of uncertainty on any conclusions that might be drawn from the time series. For each year, the square root of the error variance (the standard error) was multiplied by 1.96, and this value was added to or subtracted from the reported anomaly to define the upper and lower confidence bounds, respectively. As Figure 1 shows, the level of uncertainty has decreased dramatically in recent decades owing to better global spatial coverage and increased availability of data.

Error estimates for the gridded MLOST data set (as shown in Figure 2) have been described in a variety of articles, many of which are linked from: www.ncdc.noaa.gov/ersst/merge.php. Uncertainty measurements are also available for some of the underlying data. For example, several articles have been published about uncertainties in ICOADS in situ data; these publications are available from: www.noc.soton.ac.uk/JRD/MET/coads.php. See Box 2.1 in IPCC (2013) for additional discussion about the challenge of characterizing uncertainty in long-term climatic data sets.

11. Sources of Variability

SST varies seasonally, but Figure 1 has removed the seasonal signal by calculating annual averages. Temperatures can also vary as a result of interannual climate patterns, such as the El Niño-Southern Oscillation. Figure 2 shows how patterns in SST vary regionally.

12. Statistical/Trend Analysis

Figure 1 shows a 95 percent confidence interval that has been computed for each annual anomaly. Analysis by Smith et al. (2008) confirms that the increasing trend apparent from Figure 1 over the 20th century is statistically significant. Figure 2 shows long-term linear trends for individual grid cells on the map, and "+" symbols indicate cells where these trends are significant at a 90 percent level—an approach that is consistent with the original IPCC source map (IPCC, 2013).

References

IPCC (Intergovernmental Panel on Climate Change). 2013. Climate change 2013: The physical science basis. Working Group I contribution to the IPCC Fifth Assessment Report. Cambridge, United Kingdom: Cambridge University Press. www.ipcc.ch/report/ar5/wg1.

NOAA (National Oceanic and Atmospheric Administration). 2014a. Extended reconstructed sea surface temperature (ERSST.v3b). National Climatic Data Center. Accessed March 2014. www.ncdc.noaa.gov/ersst/.

NOAA (National Oceanic and Atmospheric Administration). 2014b. International comprehensive ocean-atmosphere data sets (ICOADS). Accessed March 2014. http://icoads.noaa.gov/.

Rayner, N.A., D.E. Parker, E.B. Horton, C.K. Folland, L.V. Alexander, D.P. Rowell, E.C. Kent, and A. Kaplan. 2003. Global analyses of sea surface temperature, sea ice, and night marine air temperature since the late nineteenth century. J. Geophys. Res. 108:4407.

Smith, T.M., and R.W. Reynolds. 2005. A global merged land-air-sea surface temperature reconstruction based on historical observations (1880–1997). J. Climate 18(12):2021-2036. www.ncdc.noaa.gov/oa/climate/research/Smith-Reynolds-dataset-2005.pdf.

Smith, T.M., R.W. Reynolds, T.C. Peterson, and J. Lawrimore. 2008. Improvements to NOAA's historical merged land-ocean surface temperature analysis (1880–2006). J. Climate 21(10):2283–2296. www.ncdc.noaa.gov/ersst/papers/SEA.temps08.pdf.

Sea Level

Identification

1. Indicator Description

This indicator describes how sea level has changed since 1880. Rising sea levels have a clear relationship to climate change through two main mechanisms: changes in the volume of ice on land (shrinking glaciers and ice sheets) and thermal expansion of the ocean as it absorbs more heat from the atmosphere. Changes in sea level are important because they can affect human activities in coastal areas and alter ecosystems.

Components of this indicator include:

- Average absolute sea level change of the world's oceans since 1880 (Figure 1)
- Trends in relative sea level change along U.S. coasts over the past half-century (Figure 2)

2. Revision History

April 2010:   Indicator posted.
December 2011:   Updated with data through 2009.
May 2012:   Updated with altimeter data through 2011 from a new source and tide gauge data from 1960 to 2011.
June 2012:   Updated with long-term reconstruction data through 2011.
August 2013:   Updated Figure 2 on EPA's website with data through 2012.
January 2014:   Updated Figure 1 with long-term reconstruction data through 2012 and altimeter data through 2013.
May 2014:   Updated Figure 2 with data through 2013.

Data Sources

3. Data Sources

Figure 1. Global Average Absolute Sea Level Change, 1880–2013

Figure 1 presents a reconstruction of absolute sea level developed by Australia's Commonwealth Scientific and Industrial Research Organisation (CSIRO). This reconstruction is available through 2012 and is based on two main data sources:

- Satellite data from the TOPography EXperiment (TOPEX)/Poseidon, Jason-1, and Jason-2 satellite altimeters, operated by the National Aeronautics and Space Administration (NASA) and France's Centre National d'Etudes Spatiales (CNES).

- Tide gauge measurements compiled by the Permanent Service for Mean Sea Level (PSMSL), which includes more than a century of daily and monthly tide gauge data.

Figure 1 also presents the National Oceanic and Atmospheric Administration's (NOAA's) analysis of altimeter data from the TOPEX/Poseidon, Jason-1 and -2, GEOSAT Follow-On (GFO), Envisat, and European Remote Sensing (ERS) 2 satellite missions. These data are available through 2013.

Figure 2. Relative Sea Level Change Along U.S. Coasts, 1960–2013

Figure 2 presents tide gauge trends calculated by NOAA. The original data come from the National Water Level Observation Network (NWLON), operated by the Center for Operational Oceanographic Products and Services (CO-OPS) within NOAA's National Ocean Service (NOS).

4. Data Availability

Figure 1. Global Average Absolute Sea Level Change, 1880–2013

The CSIRO long-term tide gauge reconstruction has been published online in graph form at: www.cmar.csiro.au/sealevel, and the data are posted on CSIRO's website at: www.cmar.csiro.au/sealevel/sl_data_cmar.html. CSIRO's website also provides a list of tide gauges that were used to develop the long-term tide gauge reconstruction.

At the time this indicator was published, CSIRO's website presented data through 2009. The same results were also published in Church and White (2011). EPA obtained an updated version of the analysis with data through 2012 from Dr. Neil White at CSIRO.

The satellite time series was obtained from NOAA's Laboratory for Satellite Altimetry, which maintains an online repository of sea level data (NOAA, 2013). The data file for this indicator was downloaded from: http://ibis.grdl.noaa.gov/SAT/SeaLevelRise/slr/slr_sla_gbl_free_all_66.csv. Underlying satellite measurements can be obtained from NASA's online database (NASA, 2014). The reconstructed tide gauge time series is based on data from the PSMSL database, which can be accessed online at: www.psmsl.org/data/.

Figure 2. Relative Sea Level Change Along U.S. Coasts, 1960–2013

The relative sea level map is based on individual station measurements that can be accessed through NOAA's "Sea Levels Online" website at: http://tidesandcurrents.noaa.gov/sltrends/sltrends.shtml. This website also presents an interactive map that illustrates sea level trends over different timeframes. NOAA has not published the specific table of 1960–2013 trends that it provided to EPA for this indicator; however, a user could reproduce these numbers from the publicly available data cited above. NOAA published an earlier version of this trend analysis in a technical report on sea level variations of the United States from 1854 to 1999 (NOAA, 2001). EPA obtained the updated 1960–2013 analysis from the lead author of NOAA (2001), Chris Zervas.

Methodology

5. Data Collection

This indicator presents absolute and relative sea level changes. Absolute sea level change (Figure 1) represents only the sea height, whereas relative sea level change (Figure 2) is defined as the change in sea height relative to land. Land surfaces move up or down in many locations around the world due to natural geologic processes (such as uplift and subsidence) and human activities that can cause ground to sink (e.g., from extraction of groundwater or hydrocarbons that supported the surface).

Sea level has traditionally been measured using tide gauges, which are mechanical measuring devices placed along the shore. These devices measure the change in sea level relative to the land surface, which means the resulting data reflect both the change in absolute sea surface height and the change in local land levels. Satellite measurement of land and sea surface heights (altimetry) began several decades ago; this technology enables measurement of changes in absolute sea level. Tide gauge data can be converted to absolute change (as in Figure 1) through a series of adjustments as described in Section 6.

The two types of sea level data (relative and absolute) complement each other, and each is useful for different purposes. Relative sea level trends show how sea level change and vertical land movement together are likely to affect coastal lands and infrastructure, while absolute sea level trends provide a more comprehensive picture of the volume of water in the world's oceans, how the volume of water is changing, and how these changes relate to other observed or predicted changes in global systems (e.g., increasing ocean heat content and melting polar ice caps). Tide gauges provide more precise local measurements, while satellite data provide more complete spatial coverage. Tide gauges are used to help calibrate satellite data. For more discussion of the advantages and limitations of each type of measurement, see Cazenave and Nerem (2004).

Tide Gauge Data

Tide gauge sampling takes place at sub-daily resolution (i.e., measured many times throughout the day) at sites around the world. Some locations have had continuous tide gauge measurements since the 1800s.

Tide gauge data for Figure 1 were collected by numerous networks of tide gauges around the world. The number of stations included in the analysis varies from year to year, ranging from fewer than 20 locations in the 1880s to more than 200 locations during the 1980s. Pre-1880 data were not included in the reconstruction because of insufficient tide gauge coverage. These measurements are documented by the PSMSL, which compiled data from various networks. The PSMSL data catalogue provides documentation for these measurements at: www.psmsl.org/data/.

Tide gauge data for Figure 2 come from NOAA's National Water Level Observation Network (NWLON). NWLON is composed of 175 long-term, continuously operating tide gauge stations along the United States coast, including the Great Lakes and islands in the Atlantic and Pacific Oceans. The map in Figure 2 shows trends for 67 stations along the ocean coasts that had sufficient data from 1960 to 2013. NOAA (2001) describes these data and how they were collected. Data collection methods are documented in a series of manuals and standards that can be accessed at: www.co-ops.nos.noaa.gov/pub.html#sltrends.

Satellite Data

Satellite altimetry has revealed that the rate of change in absolute sea level differs around the globe (Cazenave and Nerem, 2004). Factors that lead to changes in sea level include astronomical tides; variations in atmospheric pressure, wind, river discharge, ocean circulation, and water density (associated with temperature and salinity); and added or extracted water volume due to the melting of ice or changes in the storage of water on land in reservoirs and aquifers.

Data for this indicator came from the following satellite missions:

- TOPEX/Poseidon began collecting data in late 1992; Jason began to replace TOPEX/Poseidon in 2002. For more information about the TOPEX/Poseidon and Jason missions, see NASA's website at: http://sealevel.jpl.nasa.gov/missions/.

- The U.S. Navy launched GFO in 1998, and altimeter data are available from 2000 through 2006. For more information about the GFO missions, see NASA's website at: http://gcmd.nasa.gov/records/GCMD_GEOSAT_FOLLOWON.html.

- The European Space Agency (ESA) launched ERS-2 in 1995, and its sea level data are available from 1995 through 2003. More information about the mission can be found on ESA's website at: https://earth.esa.int/web/guest/missions/esa-operational-eo-missions/ers.

- ESA launched Envisat in 2002, and this indicator includes Envisat data from 2002 through 2010. More information about Envisat can be found on ESA's website at: https://earth.esa.int/web/guest/missions/esa-operational-eo-missions/envisat.

TOPEX/Poseidon and Jason satellite altimeters each cover the entire globe between 66 degrees south and 66 degrees north with 10-day resolution. Some of the other satellites have different resolutions and orbits. For example, Envisat is a polar-orbiting satellite.

6. Indicator Derivation

Satellite Data for Figure 1. Global Average Absolute Sea Level Change, 1880–2013

NOAA processed all of the satellite measurements so they could be combined into a single time series. In doing so, NOAA limited its analysis to data between 66 degrees south and 66 degrees north, which covers a large majority of the Earth's surface and represents the area with the most complete satellite coverage.

Researchers removed spurious data points. They also estimated and removed inter-satellite biases to allow for a continuous time series during the transition from TOPEX/Poseidon to Jason-1 and -2. A discussion of the methods for calibrating satellite data is available in Leuliette et al. (2004) for TOPEX/Poseidon data, and in Chambers et al. (2003) for Jason data. Also see Nerem et al. (2010).

Data were adjusted using an inverted barometer correction, which corrects for air pressure differences, along with an algorithm to remove average seasonal signals. These corrections reflect standard procedures for analyzing sea level data and are documented in the metadata for the data set. The data were not corrected for glacial isostatic adjustment (GIA)—an additional factor explained in more detail below.

NOAA provided individual measurements, spaced approximately 10 days apart (or more frequent, depending on how many satellite missions were collecting data during the same time frame). EPA generated monthly averages based on all available data points, then combined these monthly averages to determine annual averages. EPA chose to calculate annual averages from monthly averages in order to reduce the potential for biasing the annual average toward a portion of the year in which measurements were spaced more closely together (e.g., due to the launch of an additional satellite mission).

The analysis of satellite data has improved over time, which has led to a high level of confidence in the associated measurements of sea level change. Further discussion can be found in Cazenave and Nerem (2004), Miller and Douglas (2004), and Church and White (2011).

Several other groups have developed their own independent analyses of satellite altimeter data. Although all of these interpretations have appeared in the literature, EPA has chosen to include only one (NOAA) in the interest of keeping this indicator straightforward and accessible to readers. Other organizations that publish altimeter-based data include:

- The University of Colorado at Boulder: http://sealevel.colorado.edu/
- AVISO (France): www.aviso.oceanobs.com/en/data/products/ocean-indicators-products/mean-sea-level.html
- CSIRO: www.cmar.csiro.au/sealevel/

Tide Gauge Reconstruction for Figure 1. Global Average Absolute Sea Level Change, 1880–2013

CSIRO developed the long-term tide gauge reconstruction using a series of adjustments to convert relative tide gauge measurements into an absolute global mean sea level trend. Church and White (2011) describe the methods used, which include data screening; calibration with satellite altimeter data to establish patterns of spatial variability; and a correction for GIA, which represents the ongoing change in the size and shape of the ocean basins associated with changes in surface loading. On average, the world's ocean crust is sinking in response to the transfer of mass from the land to the ocean following the retreat of the continental ice sheets after the Last Glacial Maximum (approximately 20,000 years ago). Worldwide, on average, the ocean crust is sinking at a rate of approximately 0.3 mm per year. By correcting for GIA, the resulting curve actually reflects the extent to which sea level *would* be rising if the ocean basins were not becoming larger (deeper) at the same time. For more information about GIA and the value of correcting for it, see: http://sealevel.colorado.edu/content/what-glacial-isostatic-adjustment-gia-and-why-do-you-correct-it.

Seasonal signals have been removed, but no inverse barometer (air pressure) correction has been applied because a suitable long-term global air pressure data set is not available. Figure 1 shows annual average change in the form of an anomaly. EPA has labeled the graph as "cumulative sea level change" for the sake of clarity.

The tide gauge reconstruction required the use of a modeling approach to derive a global average from individual station measurements. This approach allowed the authors to incorporate data from a time-varying array of tide gauges in a consistent way. The time period for the long-term tide gauge reconstruction starts at 1880, consistent with Church and White (2011). When EPA originally published Figure 1 in 2010, this time series started at 1870. However, as Church and White have refined their methods over time, they have found the number of observations between 1870 and 1880 to be insufficient to reliably support their improved global sea level reconstruction. Thus, Church and White removed pre-1880 data from their analysis, and EPA followed suit.

Figure 2. Relative Sea Level Change Along U.S. Coasts, 1960–2013

Figure 2 shows relative sea level change for 67 tide gauges with adequate data for the period from 1960 to 2013. Sites were selected if they began recording data in 1960 or earlier and if data were available through 2013. Sites in south-central Alaska between Kodiak Island and Yakutat were excluded from the analysis because they have exhibited nonlinear behavior since a major earthquake occurred in 1964. Extensive discussion of this network and the tide gauge data analysis can be found in NOAA (2001) and in additional sources available from the CO-OPS website at: http://tidesandcurrents.noaa.gov. Generating Figure 2 involved only simple mathematics. NOAA used monthly sea level means to calculate a long-term annual rate of change for each station, which was determined by linear regression. EPA multiplied the annual rate of change by the length of the analysis period (54 years) to determine total change.

7. Quality Assurance and Quality Control

Satellite data processing involves extensive quality assurance and quality control (QA/QC) protocols—for example, to check for instrumental drift by comparing with tide gauge data (note that no instrumental drift has been detected for many years). The papers cited in Sections 5 and 6 document all such QA/QC procedures.

Church and White (2011) and earlier publications cited therein describe steps that were taken to select the highest-quality sites and correct for various sources of potential error in tide gauge measurements used for the long-term reconstruction in Figure 1. QA/QC procedures for the U.S. tide gauge data in Figure 2 are described in various publications available at: www.co-ops.nos.noaa.gov/pub.html#sltrends.

Analysis

8. Comparability Over Time and Space

Figure 1. Global Average Absolute Sea Level Change, 1880–2013

Satellite data were collected by several different satellite altimeters over different time spans. Steps have been taken to calibrate the results and remove biases over time, and NOAA made sure to restrict its analysis to the portion of the globe between 66 degrees south and 66 degrees north, where coverage is most complete.

The number of tide gauges collecting data has changed over time. The methods used to reconstruct a long-term trend, however, adjust for these changes.

The most notable difference between the two time series displayed in Figure 1 is that the long-term reconstruction includes a GIA correction, while the altimeter series does not. The uncorrected (altimeter) time series gives the truest depiction of how the surface of the ocean is changing in relation to the center of the Earth, while the corrected (long-term) time series technically shows how the volume of water in the ocean is changing. A very small portion of this volume change is not observed as absolute sea level rise (although most is) because of the GIA correction. Some degree of GIA correction is needed for a tide-gauge-based reconstruction in order to adjust for the effects of vertical crust motion.

Figure 2. Relative Sea Level Change Along U.S. Coasts, 1960–2013

Only the 67 stations with sufficient data between 1960 and 2013 were used to show sea level trends. However, tide gauge measurements at specific locations are not indicative of broader changes over space, and the network is not designed to achieve uniform spatial coverage. Rather, the gauges tend to be located at major port areas along the coast, and measurements tend to be more clustered in heavily populated areas like the Mid-Atlantic coast. Nevertheless, in many areas it is possible to see consistent patterns across numerous gauging locations—for example, rising relative sea level all along the U.S. Atlantic and Gulf Coasts.

9. Data Limitations

Factors that may impact the confidence, application, or conclusions drawn from this indicator are as follows:

1. Relative sea level trends represent a combination of absolute sea level change and local changes in land elevation. Tide gauge measurements such as those presented in Figure 2 generally cannot distinguish between these two influences without an accurate measurement of vertical land motion nearby.

2. Some changes in relative and absolute sea level may be due to multiyear cycles such as El Niño/La Niña and the Pacific Decadal Oscillation, which affect coastal ocean temperatures, salt content, winds, atmospheric pressure, and currents. The satellite data record is of insufficient length to distinguish medium-term variability from long-term change, which is why the satellite record in Figure 1 has been supplemented with a longer-term reconstruction based on tide gauge measurements.

3. Satellite data do not provide sufficient spatial resolution to resolve sea level trends for small water bodies, such as many estuaries, or for localized interests, such as a particular harbor or beach.

4. Most satellite altimeter tracks span the area from 66 degrees north latitude to 66 degrees south, so they cover about 90 percent of the ocean surface, not the entire ocean.

10. Sources of Uncertainty

Figure 1. Global Average Absolute Sea Level Change, 1880–2013

Figure 1 shows bounds of +/- one standard deviation around the long-term tide gauge reconstruction. For more information about error estimates related to the tide gauge reconstruction, see Church and White (2011).

Leuliette et al. (2004) provide a general discussion of uncertainty for satellite altimeter data. The Jason instrument currently provides an estimate of global mean sea level every 10 days, with an uncertainty of 3 to 4 millimeters.

Figure 2. Relative Sea Level Change Along U.S. Coasts, 1960–2013

Standard deviations for each station-level trend estimate were included in the data set provided to EPA by NOAA. Overall, with approximately 50 years of data, the 95 percent confidence interval around the long-term rate of change at each station is approximately +/- 0.5 mm per year. Error measurements for each tide gauge station are also described in NOAA (2001), but many of the estimates in that publication pertain to longer-term time series (i.e., the entire period of record at each station, not the 54-year period covered by this indicator).

General Discussion

Uncertainties in the data do not impact the overall conclusions. Tide gauge data do present challenges, as described by Parker (1992) and various publications available from: www.co-ops.nos.noaa.gov/pub.html#sltrends. Since 2001, there have been some disagreements and debate over the reliability of the tide gauge data and estimates of global sea level rise trends from these data (Cabanes et al., 2001). However, further research on comparisons of satellite data with tide gauge measurements, and on improved estimates of contributions to sea level rise by sources other than thermal expansion—and by Alaskan glaciers in particular—have largely resolved the question (Cazenave and Nerem, 2004; Miller and Douglas, 2004). These studies have in large part closed the gap between different methods of measuring sea level change, although further improvements are expected as more measurements and longer time series become available.

11. Sources of Variability

Changes in sea level can be influenced by multi-year cycles such as El Niño/La Niña and the Pacific Decadal Oscillation, which affect coastal ocean temperatures, salt content, winds, atmospheric pressure, and currents. The satellite data record is of insufficient length to distinguish medium-term variability from long-term change, which is why the satellite record in Figure 1 has been supplemented with a longer-term reconstruction based on tide gauge measurements.

12. Statistical/Trend Analysis

Figure 1. Global Average Absolute Sea Level Change, 1880–2013

The indicator text refers to long-term rates of change, which were calculated using ordinary least-squares regression, a commonly used method of trend analysis. The long-term tide gauge reconstruction trend reflects an average increase of 0.06 inches per year. The 1993–2012 trend is 0.12 inches per year for the reconstruction, and the 1993–2013 trend for the NOAA altimeter-based time series is 0.11 inches per year. All of these trends are highly significant statistically ($p < 0.001$). Church and White (2011) provide more information about long-term rates of change and their confidence bounds.

Figure 2. Relative Sea Level Change Along U.S. Coasts, 1960–2013

U.S. relative sea level results have been generalized over time by calculating long-term rates of change for each station using ordinary least-squares regression. No attempt was made to interpolate these data geographically.

References

Cabanes, C., A. Cazenave, and C. Le Provost. 2001. Sea level rise during past 40 years determined from satellite and in situ observations. Science 294(5543):840–842.

Cazenave, A., and R.S. Nerem. 2004. Present-day sea level change: Observations and causes. Rev. Geophys. 42(3):1–20.

Chambers, D.P., S.A. Hayes, J.C. Ries, and T.J. Urban. 2003. New TOPEX sea state bias models and their effect on global mean sea level. J. Geophys. Res. 108(C10):3-1–3-7.

Church, J.A., and N.J. White. 2011. Sea-level rise from the late 19th to the early 21st century. Surv. Geophys. 32:585–602.

Leuliette, E., R. Nerem, and G. Mitchum. 2004. Calibration of TOPEX/Poseidon and Jason altimeter data to construct a continuous record of mean sea level change. Mar. Geod. 27(1–2):79–94.

Miller, L., and B.C. Douglas. 2004. Mass and volume contributions to twentieth-century global sea level rise. Nature 428(6981):406–409. http://www.nasa.gov/pdf/121646main_2004nature.pdf.

NASA (National Aeronautics and Space Administration). 2014. Ocean surface topography from space. Accessed January 2014. http://sealevel.jpl.nasa.gov/.

Nerem, R.S., D.P. Chambers, C. Choe, and G.T. Mitchum. 2010. Estimating mean sea level change from the TOPEX and Jason altimeter missions. Mar. Geod. 33:435–446.

NOAA (National Oceanic and Atmospheric Administration). 2001. Sea level variations of the United States 1854–1999. NOAA Technical Report NOS CO-OPS 36. http://tidesandcurrents.noaa.gov/publications/techrpt36doc.pdf.

NOAA (National Oceanic and Atmospheric Administration). 2013. Laboratory for Satellite Altimetry: Sea level rise. Accessed April 2013. http://ibis.grdl.noaa.gov/SAT/SeaLevelRise/LSA_SLR_timeseries_global.php.

Parker, B.B. 1992. Sea level as an indicator of climate and global change. Mar. Technol. Soc. J. 25(4):13–24.

Land Loss Along the Atlantic Coast

Identification

1. Description

This regional feature measures the net area of undeveloped coastal land that has converted to open water since 1996, in approximately five-year increments (lustrums). Separate data series are provided for the net conversion of dry land to water, nontidal palustrine wetlands to water, and tidal wetland to water. The net conversion of land to open water is the sum of these three measures.

The submergence of coastal land is the most fundamental impact of rising sea level. Nevertheless, factors other than rising sea level can convert land to water. Conversion of dry land to wetland can also result from coastal submergence, but such conversions are not included in this regional feature.

Components of this regional feature include:

- Cumulative land area converted to open water, by region (Mid-Atlantic and Southeast) (Figure 1)
- Cumulative undeveloped land area converted to open water, by type of land lost (dry land, tidal wetland, non-tidal wetland) (Figure 2)

2. Revision History

January 2014: Feature proposed.

Data Sources

3. Data Sources

The regional feature is based on changes in land cover as mapped by the National Oceanic and Atmospheric Administration's (NOAA's) Coastal Change Analysis Program (C-CAP). This program produces a nationally standardized database of land cover and land change information for the coastal regions of the United States. C-CAP provides inventories of coastal intertidal areas, wetlands, and adjacent uplands with the goal of monitoring these habitats by updating the land cover maps every five years (see: www.csc.noaa.gov/landcover and: www.csc.noaa.gov/digitalcoast/data/ccapregional/faq). For background information about C-CAP, see Dobson et al. (1995).[1]

C-CAP has coverage for 1996, 2001, 2006, and 2011, making it possible for this feature to provide estimates of change for three lustrums. This data set is derived primarily from the data provided by the Landsat satellites, which have collected images of the Earth's surface since 1972.

---

[1] Dobson et al. (1995) provide extensive details on the original conceptualization of the C-CAP data set, but many of the details in that report do not accurately describe C-CAP as it exists today.

C-CAP is a key contributor to the federal government's Multi-Resolution Land Characteristics Consortium (MRLC), a group of federal agencies dedicated to providing digital land cover data and related information for the nation (Homer et al., 2007). The signature product of that effort is the National Land Cover Database (NLCD), which is the federal government's primary data set depicting land use and land cover in the contiguous United States. For the years 2001, 2006, and 2011, C-CAP is the source for the NLCD within the coastal zone (Vogelmann et al., 2001; Homer et al., 2007; Fry et al., 2011; Xian et al., 2011). C-CAP also has coverage for 1996.[2]

4. Data Availability

The C-CAP data set is available for the contiguous United States for the years 1996, 2001, 2006, and 2011 on NOAA's website at: www.csc.noaa.gov/digitalcoast/data/ccapregional. This site provides downloadable data files, an online viewer tool, and metadata.

Methods

5. Data Collection

This regional feature is based on the C-CAP data set. Creation of the data involves remote sensing, interpretation, and change analysis.

Remote Sensing

C-CAP is derived from the data provided by a series of Landsat satellites, which have collected imagery of the Earth's surface since 1972. Landsat is jointly managed by the U.S. Geological Survey (USGS) and the National Aeronautics and Space Administration (NASA).

As Irish (2000) explains:

> The mission of the Landsat Program is to provide repetitive acquisition of high resolution multispectral data of the Earth's surface on a global basis. Landsat represents the only source of global, calibrated, high spatial resolution measurements of the Earth's surface that can be compared to previous data records. The data from the Landsat spacecraft constitute the longest record of the Earth's continental surfaces as seen from space. It is a record unmatched in quality, detail, coverage, and value. (Irish, 2000: p.2)

The Landsat satellites have had similar orbital characteristics. For example, Landsat 7 and Landsat 8 orbit the earth 14.5 times each day, in near-polar orbits. The orbit maintains the same position relative to the sun, so as the Earth rotates, each orbital pass collects imagery to the west of the previous orbit—about 1,600 miles to the west at the equator. The "sun-synchronous" orbit means that each pass takes place between about 9:00 a.m. and 10:30 a.m., depending on latitude. Each pass can collect imagery along a path that is approximately 120 miles wide, leaving gaps between successive passes. The gaps are filled

---

[2]   Before 2001, C-CAP and the NLCD diverge. The NLCD's first (and only other) year of coverage was 1992 (Vogelmann et al., 2001). The NLCD's procedure for interpretation changed for subsequent years.

during the following days, because each day's orbits are approximately 100 miles west of the previous day's orbits (at the equator). It takes 16 days to complete the entire pattern.

The Landsat data provide multispectral imagery. That imagery is somewhat analogous to an electronic picture file: for each pixel in a picture, a digital camera measures the intensity of red, blue, and yellow light and stores these data in a file. Similarly, Landsat measures the intensity of light from different portions of the electromagnetic spectrum for each pixel, including data from both the visible and infrared wavelengths. Landsat imagery has a pixel size of approximately 30 x 30 meters.

For more information about the Landsat program, see: http://landsat.gsfc.nasa.gov/ and the resources cited therein.

Interpretation

Just as people can detect most (but not all) of what they see based on the colors their eyes perceive, geospatial analysts are able to detect whether the Earth's surface in a given 30 x 30 meter pixel is open water, vegetated wetlands, forest, grassland, bare intertidal lands (e.g., beaches and mudflats), or developed, based on the visible and infrared energy coming from a given pixel:

- Beach sand strongly reflects in all wavelengths.
- Clear, deep water strongly absorbs in all wavelengths.
- Sediment-laden water reflects visible wavelengths more than clear water, but reflects infrared wavelengths similarly to clear water.
- Vegetation strongly absorbs visible wavelengths and strongly reflects infrared.
- Marshes can be distinguished from dry land because the wet marsh soils absorb more light than the soil surface of the adjacent dry lands.

The classification of pixels based on the light emitted does not identify the land cover class everywhere. For example, tidal wetlands and sloped dry land on the southeastern side of an estuary might be difficult to distinguish with only remote sensing data: pixels may appear darker either because the land absorbs more energy due to wet soil conditions or because it is sloped toward the northwest and thus receives less light in the morning than would the identical land cover on flat ground—and it might even be in a shadow. The geospatial analysts who create the data generally obtain other information. For example:

- Information about land slopes and terrain can distinguish whether the land is sloped or flat.

- Data from the U.S. Fish and Wildlife Service's National Wetlands Inventory (NWI) also help distinguish dry land from wetlands and distinguish wetland types.

- Information on soil drainage characteristics is also useful in identifying wetland areas—for example, data from the U.S. Department of Agriculture's National Cooperative Soil Survey: http://websoilsurvey.nrcs.usda.gov/.

MRLC has developed a standardized approach for using remote sensing imagery to classify land cover (Homer et al., 2007). The C-CAP data set divides the land surface into 24 categories (NOAA, undated) which follow the MRLC classes, except where C-CAP needs more detail. Table TD-1 shows the relationship between C-CAP and NLCD data categories. C-CAP uses the Anderson et al. (1976) Level 2 classification system for dry land.

Table TD-1. Relationship Between C-CAP Land Cover Categories and Other Classification Systems

Anderson Level 1	NLCD category	C-CAP category[a]
Urban or Built-up Land (1)	Developed, High Intensity (24)	High Intensity Developed (2)
	Developed, Medium Intensity (23)	Medium Intensity Developed (3)
	Developed, Low Intensity (22)	Low Intensity Developed (4)
	Developed, Open Space (21)	Open Space Developed (5)
Agricultural Land (2)	Cultivated Crops (82)	Cultivated Land (6)
	Pasture/Hay (81)	Pasture/Hay (7)
Rangeland (3)	Grassland/Herbaceous (71)	Grassland (8)
	Scrub/Shrub (52)	Scrub Shrub (12)
Forest (4)	Deciduous Forest (41)	Deciduous Forest (9)
	Evergreen Forest (42)	Evergreen Forest (10)
	Mixed Forest (43)	Mixed Forest (11)
Wetlands (6)	Woody Wetlands (90)	Palustrine Forested Wetlands (13)
		Palustrine Scrub Shrub Wetlands (14)
		Estuarine Forested Wetlands (16)
		Estuarine Scrub Shrub Wetlands (17)
	Emergent Herbaceous Wetlands (95)	Palustrine Emergent Wetlands (15)
		Estuarine Emergent Wetlands (18)
Open Water (5)	Open Water (11)	Open Water (21)
		Palustrine Aquatic Bed (22)
		Estuarine Aquatic Bed (23)
Barren Land (7)	Barren Land (31)	Unconsolidated Shore (19)[b]
		Barren Land (20)
Tundra (8)		Tundra (24)
Perennial Ice/Snow (9)	Perennial Ice/Snow (12)	Perennial Ice/Snow (25)

a Category 1 is land that could not be otherwise classified, and does not apply to any locations along the Atlantic coast in the C-CAP data set.

b The Unconsolidated Shore class refers to lands below the daily high water mark. Dry beaches, unvegetated sand spits, and similar areas are classified as Barren Land.

Source: NOAA: www.csc.noaa.gov/digitalcoast/data/ccapregional/faq

For wetlands, C-CAP uses the more detailed categories from Cowardin et al. (1979), which are also used by the NWI. C-CAP uses NWI data to distinguish estuarine from freshwater wetlands. The NWI maps by themselves are insufficient for coastal change, because they are updated so infrequently that digital maps are only available for one or two years, which vary from state to state.[3] Nevertheless, the NWI

---

3   Along the portion of the Atlantic Coast covered by this feature, NWI maps have been updated since the year 2000 for only five states: New York (2004), Delaware (2007), North Carolina (2010), Georgia (2004–2007), and Florida south of Volusia County (2006–2010). See "Wetlands Mapper" in U.S. FWS (undated).

maps are useful for identifying the wetland class that corresponds to a given spectral signature at a given location.

Even though Landsat provides an image every 16 days, the C-CAP data are intended to represent land cover for a given year, based on data collected throughout that year—and if necessary, the preceding year. Interpretation of the Landsat data to create land cover data may require examination of many Landsat images for each location simply to find one or two that are useful. For some locations, no available imagery from a given year shows the land cover, so NOAA uses the preceding year. Distinguishing dry land and some tidal wetlands from open water requires obtaining images close to low tide. That rules out the majority of images, which are taken when the water is higher than at low tide. During winter, ice and snow cover can prevent a correct interpretation of whether a pixel is even open water or dry land, let alone the type of land. Because some types of vegetation become green sooner than others, distinguishing among those types of land cover may require imagery from a very specific time in the spring. Even when the tides and ground surfaces are perfect for measuring land cover, clouds may obscure the image on the morning that the satellite passes.

C-CAP's five-year increment represents a balance between the cost of data interpretation and the benefits of more observations.

Change Analysis

Most of the climate change indicators in this report depict change over time based on data that are measured in the same way continuously, daily, or at least once a year. While the same instruments might be used from year to year, a different measurement is taken each year, or a different annual measurement is derived from many measurements taken throughout the year. In theory, the C-CAP data could be created by following the same process once every five years, completely reinterpreting the data each time. C-CAP uses a different procedure, though, as NOAA (http://www.csc.noaa.gov/digitalcoast/data/ccapregional/faq) explains:

- Change detection analysis compares the two dates of imagery to identify the areas that have likely changed during this time frame. These areas are then classified through a combination of models, use of ancillary data, and manual edits…. The classified land cover for these areas of change is then superimposed over the land cover from the original date of analysis, to create a new wall-to-wall classification for the second time period.

- Many of the areas that are identified as potentially changed in the change detection and masking may be changes that do not result in a change in class. Agricultural field rotation or forest stand thinning are two examples.

- Remapping only areas that contain change leads to greater efficiency and more consistent data through time than would procedures that remap the full land cover for an area for each time period. By looking only at areas that have changed, we remove any difference that could result from differences in interpretation. This method also reduces cost, since only a small percentage of the total area must be classified in the change product (i.e., typically less than 20% of any area changes in a five-year period).

6. Derivation

This section describes a general approach for measuring coastal submergence with the C-CAP data. Some categories of change are more definitive than others. Although the regional feature focuses on the conversion of undeveloped land to open water, the entire approach is presented to permit a more thorough review of the methods and possible limitations of the data. For the purposes of this analysis, the Mid-Atlantic region is defined as New York, New Jersey, Delaware, Maryland, and Virginia; the Southeast includes North Carolina, South Carolina, Georgia, and the Atlantic coast of Florida (including all of Monroe County).

Simplifying the Land Classification

The first step was to group the C-CAP land classes into four general classes relevant for measuring the possible submergence of coastal land, as shown in Table TD-2: dry land, palustrine (largely nontidal, freshwater) wetlands, tidal wetlands, and open water. The developed land categories are omitted for two reasons:

1. Developed lands are generally protected from shoreline erosion and rising sea level (Titus et al., 2009), and thus rarely convert to tidal wetlands or open water.

2. Although conversion of open water or tidal wetlands into developed land is rare (and strongly discouraged by environmental regulations), the C-CAP data erroneously show it to have occurred in many locations.[4]

Given the exclusion of developed lands, this regional feature should be construed as measuring submergence of undeveloped lands.

Table TD-2. Reclassification for Evaluating Coastal Land Submergence

Land conversion category	C-CAP classification
Undeveloped Dry Land	6–12 and 20
Palustrine Wetlands	13, 14, and 15
Tidal Wetlands	16, 17, 18, 19
Open Water	21–23

Distinguishing Coastal Land Loss from Upland Changes

As a general rule, when all four general categories are present, dry land is inland and higher than nontidal (palustrine) wetlands, which are higher than tidal wetlands, which are higher than open water. Rising sea level tends to cause a higher category to convert to a lower category, and conversion from a

---

[4] When the data show a new coastal development, sometimes pixels along the shore are shown as converting from water to developed land when, in reality, the only change has been that the part of the pixel that was originally undeveloped land has converted to developed land. New docks and boats sometimes show up as new developed land. Pre-existing bridges sometimes show up as open water converting to developed land.

higher to lower category might indicate coastal submergence. Conversions from a lower to a higher category, by contrast, may indicate natural land accretion or fill activities, which generally require a permit. In addition to actual conversions, the data may show conversions in either direction due to "measurement error" resulting from the imperfections of interpreting remote sensing data.

This analysis focuses on the net conversion between categories rather than the total change in each direction, because measurement error and shoreline processes unrelated to submergence have less effect on net conversion than on the conversion in a particular direction. For example, a slight shift in the grid that defines pixel boundaries can cause apparent land gain in some places and land loss in others. Given the relative stability of most shores, these errors are more likely to cause the data to show spurious changes—in both directions—than to hide actual changes. Inlet migration and other shoreline processes may cause the shore to accrete in one place and erode nearby, and such shifts do not represent coastal submergence.

Table TD-3 summarizes the relationship between coastal submergence and the types of coastal change shown by the four land categories. Figure TD-1 illustrates two example environments. Both the table and the figure distinguish three types of conversions: those that are included in the regional feature; those that could generally be included in a land loss indicator, but are excluded from this edition of the regional feature because additional review is needed; and conversions that might result from coastal submergence, but are often caused by unrelated factors.

Conversion of land to tidal open water generally is an indicator of coastal submergence. Because remote sensing is well-suited to distinguishing land from open water, the regional feature includes conversion of land to open water, as shown in the final column of Table TD-3.

Table TD-3. Does Conversion Between Classes Generally Represent Coastal Submergence?

From↓ To→	Dry land	Palustrine[b]	Tidal wetland[c]	Open water
Dry land[a]		No	No[e]	If tidal[d]
Palustrine wetland[a]	*		No[e]	If tidal[d]
Tidal wetland[b]	*	*		Yes
Open water	*	*	*	

a Changes involving developed lands, C-CAP categories 2–5, are excluded from the feature, but would represent coastal land loss.
b "Palustrine" includes tidal freshwater wetlands.
c "Tidal wetland" includes both "estuarine wetlands" and "unconsolidated shore," but the C-CAP data set does not distinguish tidal freshwater wetlands from nontidal palustrine wetlands.
d "If tidal" means that an additional data set is used to determine whether the open water is tidal or nontidal.
e Net conversions from dry land and palustrine wetlands to tidal wetlands would generally represent net coastal submergence, but additional review of the data is necessary to ascertain whether C-CAP provides a representative measurement of such conversions.
* Treated as an offset to conversion in the other direction, when calculating net conversion.

Figure TD-1. Land Conversions Included and Excluded in this Analysis

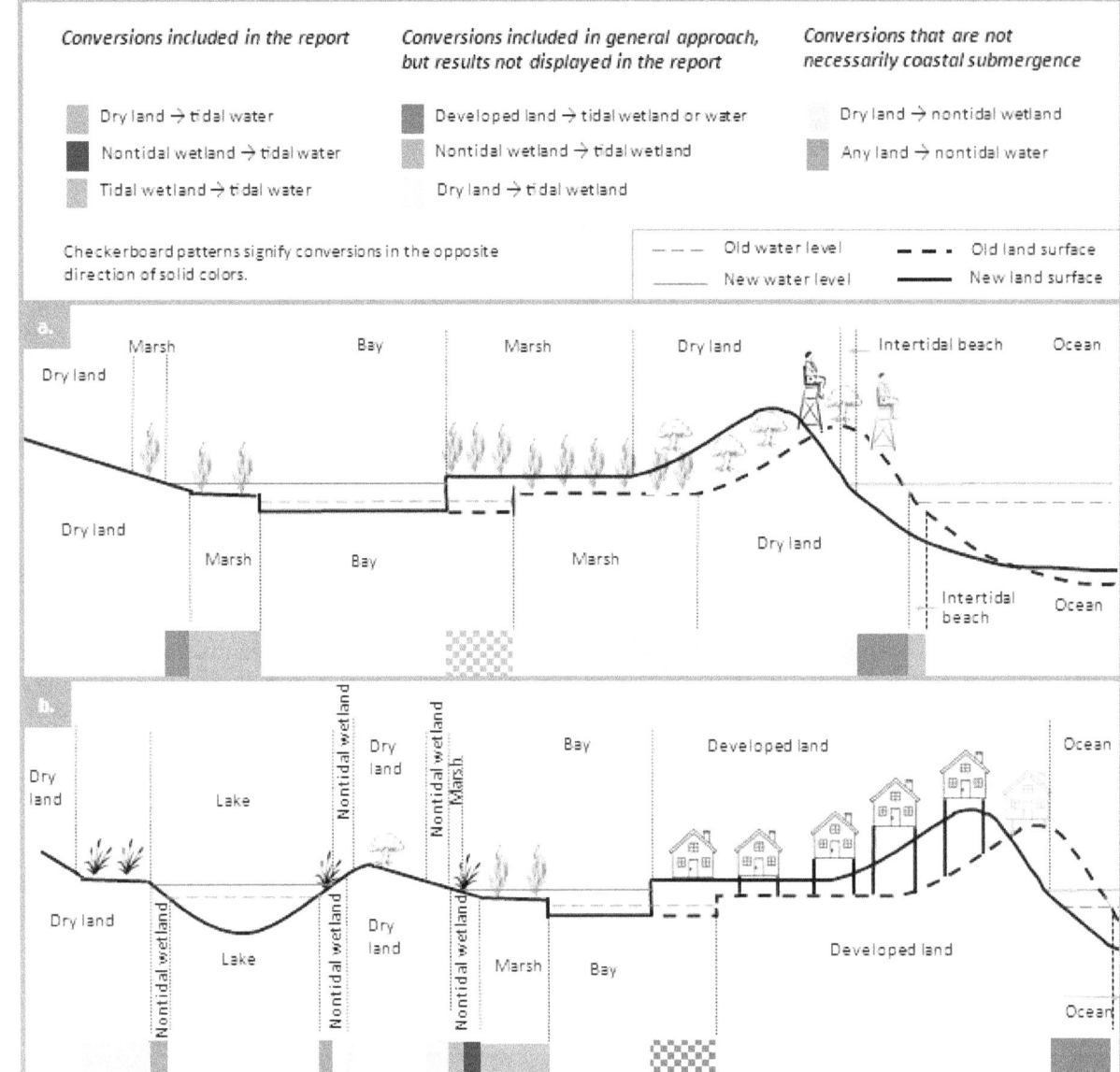

Both sketches show a barrier island that migrates inland and submergence from higher water levels along the mainland shore. (a) On the ocean side, the undeveloped dry land converts to intertidal beach, and beach to open water. The landward transport of sand, however, elevates land surfaces on the bay side, which converts some shallow water to marsh and some marsh to dry land. On the mainland shore, dry land converts to tidal wetland, and tidal wetland coverts to open water. The regional feature in the Indicators report includes the net conversions to open water, but because additional review is required, it does not include conversions between dry land and tidal wetlands. (b) The barrier island migrates as in (a), but the regional feature does not include conversions involving developed lands. Farther inland, the rising bay leads to a higher water table, which in turn saturates some soils enough to convert dry land to nontidal wetlands, and raises the level of the nearby freshwater lake. The higher lake also converts previously dry land to nontidal wetlands, and nontidal wetlands to nontidal open water. Although the conversion in this particular case might be attributed to rising sea level, in general, lake shorelines and the boundaries of nontidal wetlands change for many reasons unrelated to rising sea level. Therefore, this analysis does not treat changes in nontidal open water or conversions of dry land to nontidal wetlands as a form of coastal submergence.

Conversion of dry land or palustrine wetlands to tidal wetlands is one of the more commonly discussed consequences of sea level rise (e.g., Titus et al., 2009; Titus, 2011), and it would generally indicate coastal submergence. Additional review is necessary, however, to ensure that the C-CAP data provide a representative estimate of the net conversion of these relatively high lands to tidal wetlands.

Finally, this feature excludes net conversion of dry land to palustrine wetlands, because those conversions do not necessarily indicate coastal submergence. Rising sea level can cause low-lying dry land to convert to palustrine wetlands under a variety of circumstances. Yet most palustrine wetlands are substantially above sea level, and conversions of dry land to palustrine wetlands at substantial elevations are unrelated to coastal submergence. No multi-state data set is available to distinguish conversions caused by submergence from those that are not.

Classifying Open Water Conversions as Tidal or Nontidal

If dry land converts to *tidal* open water (e.g., beach erosion or bluff erosion), then coastal land is lost. Yet dry land might convert to *nontidal* open water if a pond is excavated for agricultural, stormwater, or industrial purposes—or if the water level in a lake happens to be higher because it was a wet year. C-CAP does not distinguish between tidal and nontidal open water, so it was necessary to use additional data sets, discussed below.

Many wetlands data sets keep track of whether open water is tidal or nontidal. Unfortunately, they are not revised with a regular five-year frequency, and many of them are relatively old. Thus, using them to distinguish coastal land loss from upland land conversion might require considerable interpretation, analogous to the effort that was necessary to distinguish tidal from nontidal wetlands in the C-CAP data set.

An alternative approach, followed here, is to use available elevation data as a proxy for whether land converts to or from tidal or nontidal open water. While the vintage of high-resolution elevation data varies, it is generally more recent than 2006 and in some cases more recent than 2011. With the exception of high bluffs in a few areas, if high-elevation land becomes open water, it is not coastal land loss, and if land of any elevation becomes high elevation water, it is virtually never coastal land loss. Similarly, it is virtually never coastal land gain when water becomes land if either the land or the water has a high elevation.

Conversely, nontidal open water is rare within 1 meter above mean high water (MHW), except for artificially managed bodies of water such as Florida's water management canals. If newly created land has an elevation within about 50 centimeters (cm) above spring high water, it is likely to be coastal land accretion; and if land below 1 meter converts to open water, it is almost always tidal open water. The same situation largely pertains when the elevation of the water surface is close to sea level, with one caveat: Elevation of water bodies based solely on lidar sometimes provides an erroneously low reading, so a different data source is necessary for elevation of water surfaces.

In light of these considerations, EPA's approach defines the coastal area vulnerable to inundation as follows:

- Include land and water within 100 meters of the open coast—including large bays and sounds—to account for shore erosion and accretion.[5]

- Include land and water less than 3 feet above mean higher high water (MHHW), based primarily on the NOAA sea level rise viewer data for land elevations, but also based on USGS 7.5-minute quadrangles where lidar elevation data are unavailable (i.e., open water areas and certain locations not yet covered by lidar). Exclude dry land, nontidal wetlands, and open water above that elevation.

- Include all tidal wetlands regardless of measured elevation.

- Exclude conversions that are clearly the result of mining or construction projects, unless the water elevations are below MHHW. Beach nourishment projects that reclaim land from the sea are treated as land gain; excavations near the shore that create open water are treated as land loss.

The 3-foot elevation cutoff was chosen to ensure that all lands within approximately 1 foot above the tidal wetland boundary are included. In areas with a large tide range, high marsh generally extends up to spring high water, which can be 1 to 2 feet above MHHW. In areas with negligible tides, high marsh can extend 1 to 2 feet above MHHW as a result of wind generated tides. The 3-foot layer would thus include all land that is within 1 to 2 feet above the tidal-wetland boundary.

To account for potential measurement error in the lidar data, the NOAA sea level rise viewer provides a confidence layer, i.e., lidar-based data sets that identify land whose measured elevation is—with a high degree of confidence—less than a given distance above MHHW. For example, the 3-foot layer (which this analysis uses) includes all land whose elevation above MHHW is less than 3 feet plus the root mean squared error of the elevation data. For further details, see Schmid et al. (undated).

NOAA's data layer does not include elevations for lakes and ponds, because lidar does not measure elevations of water surfaces. (The data layer provides values for the purposes of the viewer, but those values do not represent measured elevations.) Fortunately, other elevation data sources with less accuracy, such as USGS topographic maps, avoid this problem. Thus, EPA used a second data set to remove any locations shown to be above 10 feet in places where lidar does not provide an elevation, such as inland lakes.[6]

Results: Land Cover Change Matrices

Tables TD-4 and TD-5 are region-specific change matrices, which quantify changes between the four primary land cover categories for each of the three lustrums, along with the corresponding cumulative

---

[5] This version of the regional feature relies on visual inspection of the results to ensure that shoreline areas are not excluded, which was feasible because bluff erosion is not a major factor in the Mid-Atlantic or Southeast. This step could also be automated using a shoreline data file.

[6] This version of the regional feature relies on visual inspection of inland lakes and USGS 7.5-minute quadrangles, but it could also be automated using the USGS National Elevation Data Set.

net change. Each table provides the extent of change in square miles, which is found by dividing pixel counts by 2,878 (the number of 30 x 30 meter pixels per square mile).

The final column in each table shows the net change of undeveloped land to open water compared with the 1996 baseline, as shown in Figures 1 and 2 of this regional feature. For transparency, the tables below include the other conversion categories.

Table TD-4. Results for the Mid-Atlantic Region

Change matrix: 1996–2001 (square miles)

From↓ To→	Upland	Palus-trine	Tidal wetland	Water
Upland	1,555.90	2.19	2.22	1.28
Palus-trine	3.37	995.57	0.25	0.09
Tidal wetland	0.95	0.28	1,148.9	0.91
Water	1.44	0.73	29.23	9,165.5

Net change: 1996–2001 (square miles)

From↓ To→	Upland	Palus-trine	Tidal wetland	Water
Upland	NA	-1.18	1.27	-0.16
Palus-trine		NA	-0.02	-0.64
Tidal wetland			NA	-0.76
Water				NA
Total change	0.07	-0.51	2.00	-1.56

Change matrix: 2001–2006 (square miles)

From↓ To→	Upland	Palus-trine	Tidal wetland	Water
Upland	1,562.63	0.26	0.34	3.03
Palus-trine	10.72	999.42	0.05	2.07
Tidal wetland	5.03	0.11	1,136.58	6.56
Water	3.53	0.50	23.22	9,162.81

Net change: 1996–2006 (square miles)

From↓ To→	Upland	Palus-trine	Tidal wetland	Water
Upland	NA	-11.64	-3.42	-0.66
Palus-trine		NA	-0.08	0.92
Tidal wetland			NA	4.80
Water				NA
Total change	15.71	-12.48	-8.30	5.07

Change matrix: 2006–2011 (square miles)

From↓ To→	Upland	Palus-trine	Tidal wetland	Water
Upland	1,601.01	2.73	2.51	2.22
Palus-trine	4.59	964.30	0.14	0.52
Tidal wetland	0.56	0.03	1,132.68	1.43
Water	0.76	0.80	23.92	9,135.99

Net change: 1996–2011 (square miles)

From↓ To→	Upland	Palus-trine	Tidal wetland	Water
Upland	NA	-13.49	-1.46	0.80
Palus-trine		NA	0.03	0.65
Tidal wetland			NA	2.10
Water				NA
Total change	14.16	-14.17	-3.54	3.55

Table TD-5. Results for the Southeast Region

Change matrix: 1996–2001 (square miles)				
From↓ To→	Upland	Palus-trine	Tidal wetland	Water
Upland	3,178.58	30.65	3.48	4.97
Palus-trine	37.30	4,044.46	2.37	3.24
Tidal wetland	3.97	1.74	2,222.16	8.44
Water	3.07	2.13	155.28	11,485.13

Net change: 1996–2001 (square miles)				
From↓ To→	Upland	Palus-trine	Tidal wetland	Water
Upland	NA	-6.65	-0.49	1.91
Palus-trine		NA	0.63	1.12
Tidal wetland			NA	5.35
Water				NA
Total change	5.23	-8.40	-5.21	8.38

Change matrix: 2001–2006 (square miles)				
From↓ To→	Upland	Palus-trine	Tidal wetland	Water
Upland	3,235.55	0.38	0.71	3.89
Palus-trine	16.12	4,115.13	0.02	1.35
Tidal wetland	3.08	0.21	2,243.94	4.02
Water	1.94	0.70	150.37	11,568.98

Net change: 1996–2006 (square miles)				
From↓ To→	Upland	Palus-trine	Tidal wetland	Water
Upland	NA	-22.39	-2.86	3.87
Palus-trine		NA	0.44	1.76
Tidal wetland			NA	7.40
Water				NA
Total change	21.39	-24.59	-9.83	13.03

Change matrix: 2006–2011 (square miles)				
From↓ To→	Upland	Palus-trine	Tidal wetland	Water
Upland	3,258.45	1.52	1.75	2.92
Palus-trine	12.10	4,101.05	0.03	3.24
Tidal wetland	3.60	0.10	2,228.67	2.94
Water	1.98	1.97	150.25	11,570.33

Net change: 1996–2011 (square miles)				
From↓ To→	Upland	Palus-trine	Tidal wetland	Water
Upland	NA	-32.97	-4.71	4.81
Palus-trine		NA	0.38	3.03
Tidal wetland			NA	8.60
Water				NA
Total change	32.87	-36.38	-12.93	16.43

7. Quality Assurance and Quality Control

Thorough documentation of the quality assurance and quality control (QA/QC) methods and results is available in the technical references for the NLCD and C-CAP. Publications are available at: www.csc.noaa.gov/digitalcoast/data/ccapregional. Accuracy assessments have been conducted for the NLCD (Wickham et al., 2010, 2013) and for C-CAP (e.g., Washington Department of Ecology, 2013; NOAA, 2013).

Analysis

8. Comparability Over Time and Space

The same general data collection and analytical techniques have been applied for all of the time periods covered by this regional feature. Nevertheless, the methods are not precisely the same for all periods of time, because the C-CAP data for the year 2001 are based on remote sensing from that year, while the C-CAP data for other years are based on a combination of the remote sensing from 2001 and change analysis using 2001 as a base year.

C-CAP employs the same procedures for all locations, except for those differences that inherently result from differences in land cover. The use of ancillary wetlands data, for example, is greater in areas with the greatest amount of wetlands.

9. Data Limitations

Factors that may affect the confidence, application, or conclusions drawn from this regional feature generally fall into two categories: limitations in scope and limitations of accuracy.

1. The scope of this feature does not perfectly match the submergence of coastal land caused by changing climate. By design, the feature is under-inclusive because it omits the following types of coastal submergence:

   - Conversion of developed land to water

   - Conversion of dry land to tidal wetland

   - Conversion of dry land to nontidal wetlands resulting from higher water levels

   The feature is also over-inclusive, in that:

   - Only a portion of relative sea level rise is caused by climate change, so only a portion of land conversions from sea level rise are attributable to climate change.

   - Land can convert to tidal open water for reasons unrelated to relative sea level rise, including wave-induced erosion, invasive species or changes in hydrology that kill marsh vegetation, and construction of canals. Also, shore protection and infrastructure projects can convert water to dry land.

2. The accuracy of this feature has not been assessed, and there are reasons to expect significant error. See Section 10 for additional discussion of accuracy and possible sources of uncertainty.

10. Sources of Uncertainty

This regional feature is based on the sum of many pixels, relying on the assumption that the land or water class assigned to a given pixel applies to the entire pixel. Along the shore, however, pixels are part land and part water. Thus, the feature's accuracy is limited by both the accuracy and interpretation of individual pixels (map accuracy), and by the extent to which errors at the pixel level tend to accumulate or offset one another (bias).

Map Accuracy: Interpretation of Individual Pixels

Accuracy Assessments

Accuracy of C-CAP pixels is limited by the factors present in all land cover data sets. Accuracy assessments of the NLCD (Wickham et al., 2010, 2013) suggest that individual pixels are mapped with approximately 79 percent accuracy for all of the NLCD (Anderson Level 2) categories, and 84 percent accuracy when aggregated to the Level 1 categories (Wickham et al., 2013). These assessments do not differentiate coastal from inland areas—and they may have excluded some shorelines.[7]

Accuracy assessments of C-CAP in Washington state and the western Great Lakes found similar overall accuracy. When categories are aggregated to dry land, wetland, and open water—as this feature does— the accuracy was 95 percent and 96 percent, respectively (Washington Department of Ecology, 2013; NOAA, 2013).

The most recent accuracy assessment of the NLCD has also evaluated the accuracy with which the data capture *changes* in land cover between 2006 and 2011. Wickham et al. (2013) found that the overall accuracy of whether land cover changed was approximately 95 percent—higher than the accuracy of individual pixels, although that higher accuracy is largely driven by the fact that an overwhelming majority of pixels do not change and that fact is correctly captured by the data. Within the sample, the NLCD and the high-resolution information agreed that there was no change for 97.24 percent of the cells, and that there was a change for 1.43 percent of the cells. For 1.04 percent of the cells, however, the NLCD failed to notice actual change, while for 0.263 percent, the NLCD incorrectly found that change had occurred. Thus, 84.5 percent of the change detected by NLCD actually occurred (user accuracy), but the NLCD only picked up 57.4 percent of the actual change (producer accuracy). The higher-resolution data showed that 2.5 percent of the pixels changed, while the NLCD only shows 1.7 percent of the cells as having changed. Stehman and Wickham (2006) show, however, that this type of analysis cannot necessarily provide a statistically valid estimate of the accuracy of estimates of net change, and that the necessary accuracy assessment for all of the NLCD classes of change would be cost-prohibitive.

Land loss or land gain detected by NLCD appears to be more accurate than estimated shifts from one class of land to another. Wickham et al. (2013) found accuracies of 76 percent, 80 percent, and 93 percent when the NLCD detected land loss, land accretion, or open water remaining as water, respectively (user accuracy). NLCD is considerably less successful at detecting actual land loss and land gain (producer accuracy): when a high resolution data source detected land loss, land accretion, or water remaining as water, C-CAP matched the high-resolution data for only 21 percent, 39 percent, and 86 percent of the cells, respectively (Wickham et al. 2013, p. 301).

---

[7] To ensure that the accuracy assessment results for open water were not unduly affected by large bodies of water, large bodies of water such as Pamlico Sound, Chesapeake Bay, and the oceans were deliberately excluded. Imprecision in the shoreline files used to exclude large bodies of water may have also excluded the shore itself.

The accuracy assessments of the C-CAP change analyses did not collect sufficient data to make class-specific estimates of accuracy. Unlike the assessments of the NLCD, the C-CAP accuracy assessments generally show that C-CAP detects *more* change than higher-resolution data. For example, the Washington state assessment found a user accuracy of 45 to 50 percent and a producer accuracy of 63 to 80 percent for the case where land cover changes (Washington Department of Ecology, 2013, pp. 6–8). That study did not measure the accuracy associated with conversions between land and water.

Possible Sources of Mapping Error

No comprehensive catalogue of mapping error is available for C-CAP. The following three sources are potentially significant:

- **Tidal variations.** From lustrum to lustrum, the remote sensing imagery may be based on different positions of the tides. NOAA attempts to ensure that all data are based on a consistent position of the tides, but some variation is inevitable because the time of the tides varies. With approximately 22 observations per year and a low tide every 12½ hours, only two images per year would be taken during the hour of low tide.[8] (See discussion of remote sensing in Section 5.) Those two images may both be obscured by clouds or leaves in some locations, requiring the use of images more than 30 minutes before or after high low. Even if neither clouds nor trees obscure the image, the elevation of low tide varies greatly over the course of the year. The spring tide range (full and new moons) is often about 50 percent greater than the neap-tide range (quarter moons), and the two low tides on a given day also have different heights. Winds can also cause water levels to diverge from what would be expected based solely on the astronomic tides. Thus, even in an area where the shoreline does not change, C-CAP might show land loss for a given time period if the first image is taken at a relatively low tide while the second image was taken at a relatively high tide. This is especially the case along mudflats and beaches, where wet soils rather than vegetation lines control the zonation.

- **Conversion to tidal open water from excavation of ponds or lake-level changes.** Although remote sensing can detect the difference between land cover and open water, it cannot detect whether the open water is tidal. Instead, this regional feature relies on classifying open water based on elevation. This approach has several limitations:

    - High resolution elevation data are lacking for South Carolina and part of Florida.

    - High-resolution elevation data often mischaracterize water elevations, requiring visual inspection and/or reliance on secondary elevation data sources.

    - Nontidal open water exists at very low elevations in very flat areas.[9]

---

[8] That is, the time of day is within the hour of high tide about 8 percent of the time, and 8 percent of 22 images per year is 1.7.

[9] Including conversion of land to nontidal open water in very low-lying areas would not always be an error. For example, ponds may expand because sea level rise causes water tables to rise; or ponds are created by the mining of sand used to elevate the grade and thereby prevent inundation of other lands.

- **Conversion of land to coastal development.** C-CAP might erroneously show some cells along the shore as converting from water to land as a result of coastal development. A cell that is 50 percent water and 50 percent undeveloped land, for example, might be classified as water. If such a cell is developed as part of a larger development, the cell might be reclassified as developed land. Thus, the cell might be classified as a change from water to developed land instead of a change from undeveloped to developed land.

Accuracy of this Regional Feature

Published accuracy assessments generally focus on the accuracy of individual pixels, not on the accuracy of the resulting estimates of proportions or sums. These assessments show that land cover data contribute a substantial amount of information and are often more useful than the available alternatives. Yet mixed pixels and systematic error can impair the reliability of an indicator based on the sum of many pixels, even when they are drawn from a reasonably accurate map. The focus of existing accuracy assessments on individual pixels does not address whether or not limitations in the data undermine the usefulness of the data for creating an indicator.

Methods have been developed and applied for assessing estimates of net change (Stehman and Wickham, 2006), but they have not been applied to C-CAP results. An accuracy assessment of this regional feature is necessary before confidence in it can be high. Comparison with an independent data set such as that used for the NWI Status and Trends reports (e.g. Dahl, 2006; Dahl, 2011; Dahl and Stedman, 2013) would be appropriate.

Mixed Pixels

Like most quantifications of land cover, this analysis makes the fundamental oversimplification that land cover is uniform within each cell. Challenges related to this "mixed pixel" problem (Fisher, 1997) can be safely disregarded if areas with homogenous land cover are large relative to the cell size. Conversely, in constructing an indicator of shoreline migration, the implications of mixed pixels must be part of the analysis because the shoreline consists of mixed pixels. The fact that shorelines generally change very little compared with the size of a pixel affects both accuracy at the pixel level and the accuracy of an indicator or feature based on the sum of pixel-level changes.

Over the course of five or even 15 years, most shorelines change by a small fraction of the 30-meter pixel width. Thus, a land-loss indicator based on the sum of many 30-meter pixels will only be useful if the pixels where the change is too small to be detected (and thus assumed to be zero) are roughly offset by pixels where the change is detected (and thus assumed to be 900 m^2). The error introduced by assuming that the entire mixed pixel is either water or land is a type of "scale error;" that is, the error becomes less, and eventually negligible, as pixel size decreases.[10]

Figure TD-2 illustrates the general case of small shoreline change relative to pixel size. The figure assumes that pixels with a majority of water are classified as water. While that threshold is usually

---

[10] Scale error is the rounding error necessitated by classifying a mixed pixel as one category or the other. One might define classification error as follows: $e_i = X_i - C_i$, where X is the true area of water and C is either 0 or 900 m^2. If there is no measurement error (i.e., $|e_i| < 450$ m^2 so that measurement error has no impact on C_i), then all error is scale error.

higher,[11] the same concepts apply regardless of the threshold. Initially, seven of the 15 pixels along the shore are more than 50 percent water, and the shoreline is about the length of 12 pixels. During a given time period, the shore retreats by about 1/6 of the width of a pixel, so the total land loss is about 2 pixels. Two pixels that had been entirely dry land become partly open water, and one pixel loses the small amount of land it originally had. At the end of the time period, the majority of the pixels labeled "2" and "3" are water; pixel "1" is approximately 50 percent water as well. With the assumed classification procedure, two or three pixels convert from land to water, which reasonably reflects the actual land loss of two pixels. The other 14 pixels where the land loss is undetected and rounded to zero are offset by these two or three pixels where land loss is detected and assumed to be the entire pixel.

Figure TD-2. Illustration of the "Mixed Pixel" Problem

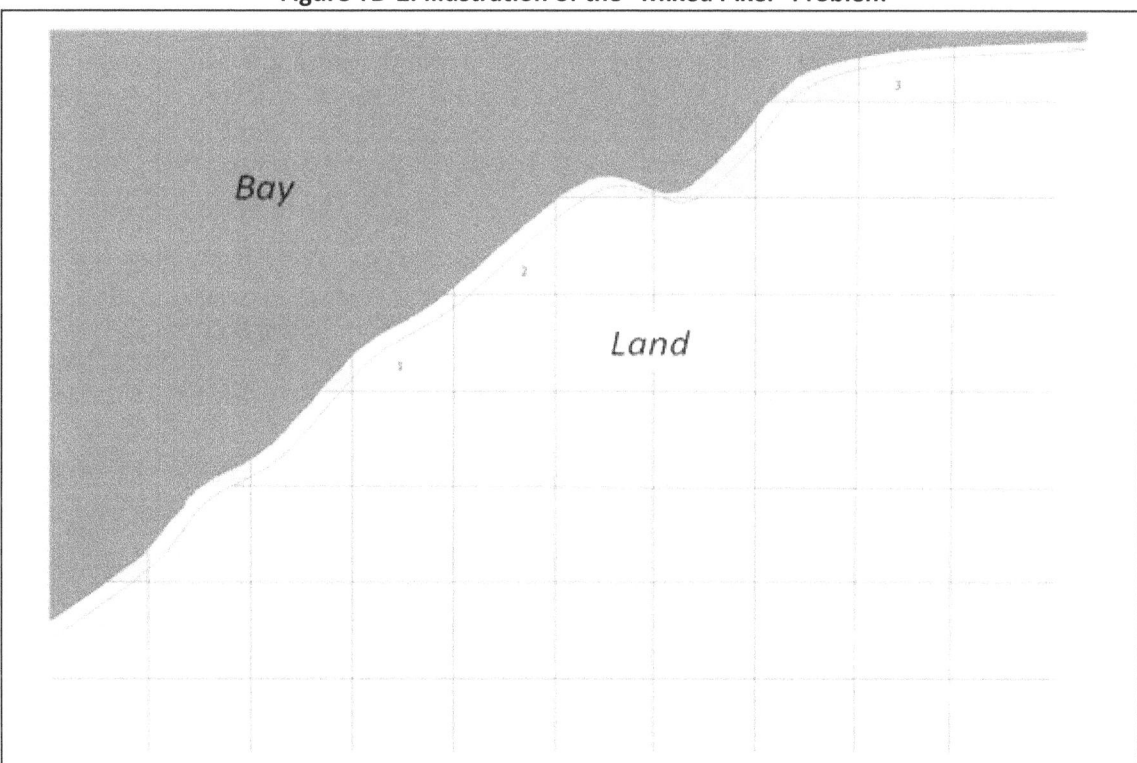

This figure shows the impact of the "mixed pixel" problem when shoreline retreats a small fraction of the width of a pixel. In this conceptual map, dark blue represents open water and white represents land. Light blue shows the dry land portion of pixels classified as water at the beginning of the time period. Cells 1, 2, and 3 might be classified as open water after the shore erodes.

Scale error does not necessarily compromise the validity of an indicator based on the sum of all pixels. The grid that defines the pixel boundaries is set independently of the shoreline, so the pixel-specific

---

[11] In general, a pixel is classified as "water" if the spectral signature is more consistent with water than with land. The threshold for classifying a pixel as water varies with locations. The NLCD has consistently defined open water as areas with open water, "generally with less than 25% cover of vegetation or soil." See MRLC (undated) and Wickham et al. (2013), p. 295.

scale error is random with a mean of zero.[12] Thus, while the scale error can be large at the scale of a pixel, it should be very small for a shoreline 100 kilometers long.[13] Random measurement errors also tend to cancel and make a negligible contribution to a total.

Systematic Error

Measurement error is not always random. Systematic measurement errors tend to accumulate rather than cancel out as one calculates the total (e.g., see Cochran, 1977). Nevertheless, a systematic error concerning the *area* of water in a sum of mixed pixels does not necessarily provide a poor estimate of the *change* in the area of water (i.e., land loss). For example, if it were shown that C-CAP classifies cells with at least *H* percent water as being water, and *H* varies but is on the order of 20 or 30 percent, then C-CAP would overestimate the area of water among the mixed pixels in possibly unpredictable ways. Yet the estimated change could still be reasonably accurate,[14] as long as the procedure overestimates the area of water consistently.

---

[12] The scale error for the mixed pixels would have a rectangular distribution from −450 m^2 to +450 m^2. Given the well-established moments of a rectangular distribution, the mean scale error is zero and σ_e^2 = 67,500 m^4; i.e., σ_e=260 m^2.

[13] If scale error is uncorrelated across pixels and measurement error is negligible, the variance of the sum of pixels would be $\sigma_S^2 = \sum_i^N \sigma_e^2$ = N 67,500 m^4. If the shoreline is 100 km, $N \approx 3,333$ so $\sigma_S^2 \approx$ 225,000,000 m^4 and $\sigma_S \approx$ 0.015 km^2. As Figure B shows, the scale error between adjacent cells is correlated, but that correlation becomes negligible more than five to 10 cells out. Conservatively assuming perfect correlation between the 10 adjacent cells, the standard deviation would be a factor of 10 greater, which is still negligible for a shoreline of 100 km or longer.

[14] For example, suppose that the probability that a pixel will be classified is an unknown function of X_i:

prob(class=water) $\equiv$ prob(C_i=1) = F(X_i), with F(0)=0 and F(1)=1.

In this note, all areas are expressed in pixels rather than m^2. Among the mixed pixels, X_i is uniformly distributed between 0 and 1, so the expected value of the sum of pixels S_1 would be:

$$E(S_1) = N_w + N_m \int_0^1 F(y)dy,$$

where N_w and N_m are the number of all-water and mixed pixels, respectively. In the ideal case with no measurement error where F is a step function

$$F(X) = 0 \text{ for } X < 0.5 \text{ and } F(X) = 1 \text{ for } X \geq 0.5, E(S_1) = N_w + N_m /2.$$

Suppose shores erode by a consistent amount and convert an area of land equal to N_m z pixels to open water, where z is a small fraction (e.g., shores erode by approximately 30z meters, a small fraction z of the pixel dimension). The amount of water increases in every pixel by z, except for those pixels for which 1-X_i<z. Given the uniform distributions, z is the fraction of pixels for which this occurs; that is, zN_m mixed pixels become all water. Those zN_m pixels lose less than z because they have less than z to give, but as Figure TD-2 shows, there are adjacent pixels that previously were all land, which each lose the land that the zN_m pixels would have lost had they been farther landward with more land to give. At the end of the lustrum, there must still be a uniform probability distribution of the values of X for the same reason the distribution was uniform initially. Over long periods of time, the total shoreline may change—for example, as islands disappear. Yet in the short run, the shoreline and hence the total number of mixed pixels is constant. Therefore, the pixels that convert from land to mixed must offset those that convert from mixed to all water, and

$$E(S_2) = N_w + zN_m + N_m \int_0^1 F(y)dy, \text{ so}$$
$$E(S_2-S_1) = zN_m,$$

which would make the land loss indicator unbiased with a relatively small variance similar to σ_S^2, though possibly twice as great if errors from period to period are random, and less if some persist.

The consistency of C-CAP's measurement error from period to period is not known. The procedures used to classify land cover during the base year differ from the procedures used during subsequent years, so it is possible that the measurement error is not consistent from year to year. Significant classification errors are possible: The tendency for the change analysis to discount small changes could mean that an indicator based on a sum of pixels does not fully reflect the land loss or gain caused by small changes.[15] Moreover, classification errors may be biased or highly correlated across cells, but in different ways for different years. Along beaches and mudflats, the classifications of land and water are sensitive to whether water levels in estuaries were atypically low or high when the imagery was taken. Errors may be correlated across pixels because a single Landsat pass may obtain imagery for a large section of coast; if estuarine water levels are high or low due to rainfall, such conditions may persist and apply elsewhere. Interpretation errors may also be systematic to the extent that they result from the consistent application of an imperfect procedure.

11. Sources of Variability

As with the other time series indicators and features in this report, estimates of change from period to period have far more uncertainty than estimates over the entire period. Coastal erosion in a given region may be episodic, resulting from storms or an unusually high rate of sea level rise over a given five-year period. Rates of sea level rise may also be greater during one time period than during the next. The geography of the U.S. coast is highly variable, and there is no reason to expect land loss or gain at one location to be similar to land loss or gain elsewhere.

12. Statistical/Trend Analysis

The data in this feature have not been analyzed to determine whether they reflect statistically significant changes over time.

References

Anderson, J.R., E.E. Hardy, J.T. Roach, and R.E. Witmer. 1976. A land use and land cover classification system for use with remote sensor data. Geological Survey Professional Paper 964. U.S. Geological Survey. http://landcover.usgs.gov/pdf/anderson.pdf

Cochran, W.G. 1977. Sampling Techniques. New York. Wiley and Sons.

Cowardin, L.M., V. Carter, F.C. Golet, and E.T. LaRoe. 1979. Classification of wetlands and deepwater habitats of the United States. U.S. Fish and Wildlife Service. www.fws.gov/wetlands/Documents/Classification-of-Wetlands-and-Deepwater-Habitats-of-the-United-States.pdf.

Dobson J.E., E.A. Bright, R.L. Ferguson, D.W. Field, L.L. Wood, K.D. Haddad, H. Iredale III, J.R. Jenson, V.V. Klemas, R.J. Orth, and J.P. Thomas. 1995. Coastal Change Analysis Program (C-CAP): Guidance for

---

[15] The key assumption in the previous footnote is that the same, possibly flawed, F(X) classification applies for both periods of time. If different procedures are applied in different years, then our demonstration that the indicator should be unbiased no longer applies.

regional implementation. NOAA Technical Report, Department of Commerce, NMFS 123.
http://spo.nwr.noaa.gov/tr123.pdf.

Fisher, P. 1997. The pixel: A snare and a delusion. Int. J. Remote Sens. 18:679–685.

Fry, J.A., G. Xian, S. Jin, J.A. Dewitz, C.G. Homer, L. Yang, C.A. Barnes, N.D. Herold, and J. Wickham. 2011. Completion of the 2006 National Land Cover Database for the conterminous United States. Photogramm. Eng. Rem. S. 77:858–864.

Homer, C., J. Dewitz, M. Coan, N. Hossain, C. Larson, N. Herold, A. McKerrow, N. VanDriel, and J. Wickham. 2007. Completion of the 2001 National Land Cover Database for the conterminous United States. Photogramm. Eng. Rem. S. 73:337–341.

Irish, R.R. 2000. Landsat 7 science data user's handbook. Report 430-15-01-003-0. National Aeronautics and Space Administration.

MRLC (Multi-Resolution Land Characteristics Consortium). Undated. National Land Cover Dataset 1992: Product legend. U.S. Geological Survey. Accessed April 1, 2014. www.mrlc.gov/nlcd92_leg.php.

NOAA (National Oceanic and Atmospheric Administration). 2013. Western Great Lakes 2010 Coastal Change Analysis Program Accuracy Assessment. Charleston, SC: Coastal Services Center.

Schmid, K., B. Hadley, and K. Waters. Undated. Mapping and portraying inundation uncertainty of bathtub-type models. J. Coastal Res. (in press).

Smith, L.C. 1997. Satellite remote sensing of river inundation area, stage, and discharge: A review. Hydrol. Process. (11):1427–1439.

Stehman, S. V. and J. D. Wickham. 2006. Assessing accuracy of net change derived from land cover maps. Photogramm. Eng. Rem. S. 72(2):175–186.

Titus, J.G. 2011. Rolling easements. Washington, D.C. U.S. EPA. EPA-430-R-11-001.
http://water.epa.gov/type/oceb/cre/upload/rollingeasementsprimer.pdf.

Titus, J.G., E.K. Anderson, D.R. Cahoon, S. Gill, R.E. Thieler, and J.S. Williams. 2009. Coastal sensitivity to sea-level rise: A focus on the Mid-Atlantic region. U.S. Climate Change Science Program and the Subcommittee on Global Change Research. http://downloads.globalchange.gov/sap/sap4-1/sap4-1-final-report-all.pdf.

Titus J.G., and J. Wang. 2008. Maps of lands close to sea level along the Middle Atlantic coast of the United States: An elevation data set to use while waiting for LIDAR. Section 1.1 in: Titus, J.G., and E.M. Strange (eds.). Background documents supporting Climate Change Science Program Synthesis and Assessment Product 4.1, U.S. EPA. EPA 430R07004.
http://papers.risingsea.net/federal_reports/Titus_and_Strange_EPA_section1_1_Titus_and_Wang_may_2008.pdf.

U.S. FWS (U.S. Fish and Wildlife Service). Undated. National Wetlands Inventory.
www.fws.gov/Wetlands.

Vogelmann, J. E., S.M. Howard, L. Yang, C.R. Larson, K.K. Wylie, and J.N. Van Driel. 2001. Completion of the 1990s National Land Cover Dataset for the conterminous United States. Photogramm. Eng. Rem. S. 67:650–652.

Vogelmann, J.E., T.L. Sohl, and S.M. Howard. 1998. Regional characterization of land cover using multiple sources of data. Photogramm. Eng. Rem. S. 64:45–57.

Vogelmann, J.E., T.L. Sohl, P. V. Campbell, and D.M. Shaw. 1998. Regional land cover characterization using Landsat Thematic Mapper data and ancillary data sources. Environ. Monit. Assess. 451:415–428.

Washington Department of Ecology. 2013. Assessment report of wetland mapping improvements to NOAA's Coastal Change Analysis Program (C-CAP) land cover in western Washington state. Olympia, WA: State of Washington Department of Ecology. http://www.ecy.wa.gov/programs/sea/wetlands/pdf/C-CAPWetlandAssessmentReport.pdf.

Wickham, J.D., S.V. Stehman, L. Gass, J. Dewitz, J.A. Fry, and T.G. Wade. 2013. Accuracy assessment of NLCD 2006 land cover and impervious surface. Remote Sens. Environ. 130:294–304.

Wickham, J.D., S.V. Stehman, J.A. Fry, J.H. Smith, and C.G. Homer. 2010. Thematic accuracy of the NLCD 2001 land cover for the conterminous United States. Remote Sens. Environ. 114:1286–1296.

Wickham, J., C. Homer, J. Fry, J. Vogelmann, A. McKerrow, R. Mueller, N. Herold, J. Coulston. In review. The Multi-Resolution Land Characteristics (MRLC) Consortium - 20 years of development and integration of U.S. national land cover data. Unpublished manuscript under review.

Xian, G., C. Homer, J. Dewitz, J. Fry, N. Hossain, and J. Wickham. 2011. Change of impervious surface area between 2001 and 2006 in the conterminous United States. Photogramm. Eng. Rem. S. 77:758–762.

Ocean Acidity

Identification

1. Indicator Description

This indicator shows recent trends in acidity levels in the ocean at three key locations. The indicator also presents changes in aragonite saturation by comparing historical data with the most recent decade. Ocean acidity and aragonite saturation levels are strongly affected by the amount of carbon dissolved in the water, which is directly related to the amount of carbon dioxide (CO_2) in the atmosphere. Acidity affects the ability of corals, some types of plankton, and other creatures to produce their hard skeletons and shells. This indicator provides important information about an ecologically relevant effect associated with climate change.

Components of this indicator include:

- Recent trends in ocean CO_2 and acidity levels (Figure 1)
- Historical changes in the aragonite saturation of the world's oceans (Figure 2)

2. Revision History

April 2010:   Indicator posted.
May 2012:   Updated Figure 1 data; new Figure 2 source and metric.
May 2014:   Updated Figure 1 with data through 2012 for two sampling locations; updated Figure 2 with trends through 2013.

Data Sources

3. Data Sources

Figure 1 includes trend lines from three different ocean time series: the Bermuda Atlantic Time-Series Study (BATS); the European Station for Time-Series in the Ocean, Canary Islands (ESTOC); and the Hawaii Ocean Time-Series (HOT).

Figure 2 contains aragonite saturation (Ω_{ar}) calculations derived from atmospheric CO_2 records from ice cores and observed atmospheric concentrations at Mauna Loa, Hawaii. These atmospheric CO_2 measurements are fed into the Community Earth Systems Model (CESM), maintained by the National Center for Atmospheric Research (NCAR). CESM is a dynamic ocean model that computes ocean CO_2 uptake and the resulting changes in seawater carbonate ion (CO_3^{2-}) concentration and Ω_{ar} over time.

4. Data Availability

Figure 1 compiles pCO_2 (the mean seawater CO_2 partial pressure in μatm) and pH data from three sampling programs in the Atlantic and Pacific Oceans. Raw data from the three ocean sampling programs are publicly available online. In the case of Bermuda and the Canary Islands, updated data were procured directly from the scientists leading those programs. BATS data and descriptions are available at: http://bats.bios.edu/bats_form_bottle.html. ESTOC data can be downloaded from: www.eurosites.info/estoc/data.php. HOT data were downloaded from the HOT Data Organization and Graphical System website at: http://hahana.soest.hawaii.edu/hot/products/products.html. Additionally, annual HOT data reports are available at: http://hahana.soest.hawaii.edu/hot/reports/reports.html.

The map in Figure 2 is derived from the same source data as NOAA's Ocean Acidification "Science on a Sphere" video simulation at: http://sos.noaa.gov/Datasets/list.php?category=Ocean (Feely et al., 2009). EPA obtained the map data from Dr. Ivan Lima of the Woods Hole Oceanographic Institution (WHOI).

Methodology

5. Data Collection

Figure 1. Ocean Carbon Dioxide Levels and Acidity, 1983–2012

This indicator reports on the pH of the upper 5 meters of the ocean and the corresponding partial pressure of dissolved CO_2 (pCO_2). Each data set covers a different time period:

- BATS data used in this indicator are available from 1983 to 2012. Samples were collected from two locations in the Atlantic Ocean near Bermuda (BATS and Hydrostation S, at 31°43' N, 64°10' W and 32°10' N, 64°30' W, respectively). See: http://bats.bios.edu/bats_location.html.

- ESTOC data are available from 1995 to 2009. ESTOC is at (29°10' N, 15°30' W) in the Atlantic Ocean.

- HOT data are available from 1988 to 2012. The HOT station is at (23° N, 158° W) in the Pacific Ocean.

At the BATS and HOT stations, dissolved inorganic carbon (DIC) and total alkalinity (TA) were measured directly from water samples. DIC accounts for the carbonate and bicarbonate ions that occur when CO_2 dissolves to form carbonic acid, while total alkalinity measures the buffering capacity of the water, which affects the partitioning of DIC among carbonate and bicarbonate ions. At ESTOC, pH and alkalinity were measured directly (Bindoff et al., 2007).

Each station followed internally consistent sampling protocols over time. Bates et al. (2012) describe the sampling plan for BATS. Further information on BATS sampling methods is available at: http://bats.bios.edu. ESTOC sampling procedures are described by González-Dávila et al. (2010). HOT sampling procedures are described in documentation available at: http://hahana.soest.hawaii.edu/hot/hot_jgofs.html and: http://hahana.soest.hawaii.edu/hot/products/HOT_surface_CO2_readme.pdf.

Figure 2. Changes in Aragonite Saturation of the World's Oceans, 1880–2013

The map in Figure 2 shows the estimated change in sea surface Ω_{ar} from 1880 to 2013. Aragonite saturation values are calculated in a multi-step process that originates from historical atmospheric CO_2 concentrations that are built into the model (the CESM). As documented in Orr et al. (2001), this model uses historical atmospheric CO_2 concentrations based on ice cores and atmospheric measurements (the latter collected at Mauna Loa, Hawaii).

6. Indicator Derivation

Figure 1. Ocean Carbon Dioxide Levels and Acidity, 1983–2012

At BATS and HOT stations, pH and pCO_2 values were calculated based on DIC and TA measurements from water samples. BATS analytical procedures are described by Bates et al. (2012). HOT analytical procedures are described in documentation available at: http://hahana.soest.hawaii.edu/hot/hot_jgofs.html and: http://hahana.soest.hawaii.edu/hot/products/HOT_surface_CO2_readme.pdf. At ESTOC, pCO_2 was calculated from direct measurements of pH and alkalinity. ESTOC analytical procedures are described by González-Dávila et al. (2010). For all three locations, Figure 1 shows in situ measured or calculated values for pCO_2 and pH, as opposed to values adjusted to a standard temperature.

The lines in Figure 1 connect points that represent individual sampling events. No attempt was made to generalize data spatially or to portray data beyond the time period when measurements were made. Unlike some figures in the published source studies, the data shown in Figure 1 are not adjusted for seasonal variability. The time between sampling events is somewhat irregular at all three locations, so moving averages and monthly or annual averages based on these data could be misleading. Thus, EPA elected to show individual measurements in Figure 1.

Figure 2. Changes in Aragonite Saturation of the World's Oceans, 1880–2013

The map in Figure 2 was developed by WHOI using the CESM, which is available publicly at: www2.cesm.ucar.edu/models. Atmospheric CO_2 concentrations were fed into the CESM, which is a dynamic ocean model that computes ocean CO_2 uptake and the resulting changes in seawater carbonate concentration over time. The CESM combines this information with monthly salinity and temperature data on an approximately 1° by 1° grid. Next, these monthly model outputs were used to approximate concentrations of the calcium ion (Ca^{2+}) as a function of salt (Millero, 1982), and to calculate aragonite solubility according to Mucci (1983). The resulting aragonite saturation state was calculated using a standard polynomial solver for MATLAB, which was developed by Dr. Richard Zeebe of the University of Hawaii. This solver is available at: www.soest.hawaii.edu/oceanography/faculty/zeebe_files/CO2_System_in_Seawater/csys.html.

Aragonite saturation state is represented as Ω_{ar}, which is defined as:

$$\Omega_{ar} = [Ca^{2+}][CO_3^{2-}] / K'_{sp}$$

The numerator represents the product of the observed concentrations of calcium and carbonate ions. K'_{sp} is the apparent solubility product, which is a constant that is equal to $[Ca^{2+}][CO_3^{2-}]$ at equilibrium for a given set of temperature, pressure, and salinity conditions. Thus, Ω_{ar} is a unitless ratio that compares the observed concentrations of calcium and carbonate ions dissolved in the water with the concentrations that would be observed under fully saturated conditions. An Ω_{ar} value of 1 represents full saturation, while a value of 0 indicates that no calcium carbonate is dissolved in the water. Ocean water at the surface can be supersaturated with aragonite, however, so it is possible to have an Ω_{ar}

value greater than 1, and it is also possible to experience a decrease over time, yet still have water that is supersaturated.

For Figure 2, monthly model outputs were averaged by decade before calculating Ω_{ar} for each grid cell. The resulting map is based on averages for two decades: 1880 to 1889 (a baseline) and 2004 to 2013 (the most recent complete 10-year period). Figure 2 shows the change in Ω_{ar} between the earliest (baseline) decade and the most recent decade. It is essentially an endpoint-to-endpoint comparison, but using decadal averages instead of individual years offers some protection against inherent year-to-year variability. The map has approximately 1° by 1° resolution.

7. Quality Assurance and Quality Control

Quality assurance and quality control (QA/QC) steps are followed during data collection and data analysis. These procedures are described in the documentation listed in Sections 5 and 6.

Analysis

8. Comparability Over Time and Space

Figure 1. Ocean Carbon Dioxide Levels and Acidity, 1983–2012

BATS, ESTOC, and HOT each use different methods to determine pH and pCO_2, though each individual sampling program uses well-established methods that are consistent over time.

Figure 2. Changes in Aragonite Saturation of the World's Oceans, 1880–2013

The CESM calculates data for all points in the Earth's oceans using comparable methods. Atmospheric CO_2 concentration values differ in origin depending on their age (i.e., older values from ice cores and more recent values from direct atmospheric measurement). However, all biogeochemical calculations performed by the CESM use the atmospheric CO_2 values in the same manner.

9. Data Limitations

Factors that may impact the confidence, application, or conclusions drawn from this indicator are as follows:

1. Carbon variability exists in the surface layers of the ocean as a result of biological differences, changing surface temperatures, mixing of layers as a result of ocean circulation, and other seasonal variations.

2. Changes in ocean pH and mineral saturation caused by the uptake of atmospheric CO_2 can take a long time to spread to deeper waters, so the full effect of atmospheric CO_2 concentrations on ocean pH may not be seen for many decades, if not centuries.

3. Ocean chemistry is not uniform throughout the world's oceans, so local conditions could cause a pH measurement to seem incorrect or abnormal in the context of the global data. Figure 1 is limited to three monitoring sites.

4. Although closely tied to atmospheric concentrations of CO_2, aragonite saturation is not exclusively controlled by atmospheric CO_2, as salinity and temperature are also factored into the calculation.

10. Sources of Uncertainty

Figure 1. Ocean Carbon Dioxide Levels and Acidity, 1983–2012

Uncertainty measurements can be made for raw data as well as analyzed trends. Details on uncertainty measurements can be found in the following documents and references therein: Bindoff et al. (2007), Bates et al. (2012), Dore et al. (2009), and González-Dávila et al. (2010).

Figure 2. Changes in Aragonite Saturation of the World's Oceans, 1880–2013

Uncertainty and confidence for CESM calculations, as they compare with real-world observations, are measured and analyzed in Doney et al. (2009) and Long et al. (2013). Uncertainty for the approximation of Ca^{2+} and aragonite solubility are documented in Millero (1982) and Mucci (1983), respectively.

11. Sources of Variability

Aragonite saturation, pH, and pCO_2 are properties of seawater that vary with temperature and salinity. Therefore, these parameters naturally vary over space and time. Variability in ocean surface pH and pCO_2 data has been associated with regional changes in the natural carbon cycle influenced by changes in ocean circulation, climate variability (seasonal changes), and biological activity (Bindoff et al., 2007).

Figure 1. Ocean Carbon Dioxide Levels and Acidity, 1983–2012

Variability associated with seasonal signals is still present in the data presented in Figure 1. This seasonal variability can be identified by the oscillating line that connects sampling events for each site.

Figure 2. Changes in Aragonite Saturation of the World's Oceans, 1880–2013

Figure 2 shows how changes in Ω_{ar} vary geographically. Monthly and yearly variations in CO_2 concentrations, temperature, salinity, and other relevant parameters have been addressed by calculating decadal averages.

12. Statistical/Trend Analysis

This indicator does not report on the slope of the apparent trends in ocean acidity and pCO_2 in Figure 1. The long-term trends in Figure 2 are based on an endpoint-to-endpoint comparison between the first decade of widespread data (the 1880s) and the most recent complete 10-year period (2004–2013). The statistical significance of these trends has not been calculated.

References

Bates, N.R., M.H.P. Best, K. Neely, R. Garley, A.G. Dickson, and R.J. Johnson. 2012. Detecting anthropogenic carbon dioxide uptake and ocean acidification in the North Atlantic Ocean. Biogeosciences 9:2509–2522.

Bindoff, N.L., J. Willebrand, V. Artale, A, Cazenave, J. Gregory, S. Gulev, K. Hanawa, C. Le Quéré, S. Levitus, Y. Nojiri, C.K. Shum, L.D. Talley, and A. Unnikrishnan. 2007. Observations: Oceanic climate change and sea level. In: Climate change 2007: The physical science basis (Fourth Assessment Report). Cambridge, United Kingdom: Cambridge University Press.

Doney, S.C., I. Lima, J.K. Moore, K. Lindsay, M.J. Behrenfeld, T.K. Westberry, N. Mahowald, D.M. Glober, and T. Takahashi. 2009. Skill metrics for confronting global upper ocean ecosystem-biogeochemistry models against field and remote sensing data. J. Marine Syst. 76(1–2):95–112.

Dore, J.E., R. Lukas, D.W. Sadler, M.J. Church, and D.M. Karl. 2009. Physical and biogeochemical modulation of ocean acidification in the central North Pacific. P. Natl. Acad. Sci. USA 106:12235–12240.

Feely, R.A., S.C. Doney, and S.R. Cooley. 2009. Ocean acidification: Present conditions and future changes in a high-CO_2 world. Oceanography 22(4):36–47.

González-Dávila, M., J.M. Santana-Casiano, M.J. Rueda, and O. Llinás. 2010. The water column distribution of carbonate system variables at the ESTOC site from 1995 to 2004. Biogeosciences 7:1995–2032.

Long, M.C., K. Lindsay, S. Peacock, J.K. Moore, and S.C. Doney. 2013. Twentieth-century ocean carbon uptake and storage in CESM1(BGC). J. Climate 26(18):6775-6800.

Millero, F.J. 1982. The thermodynamics of seawater. Part I: The PVT properties. Ocean Phys. Eng. 7(4):403–460.

Mucci, A. 1983. The solubility of calcite and aragonite in seawater at various salinities, temperatures, and one atmosphere total pressure. Am. J. Sci. 283:780–799.

Orr, J.C., E. Maier-Reimer, U. Mikolajewicz, P. Monfray, J.L. Sarmiento, J.R. Toggweiler, N.K. Taylor, J. Palmer, N. Gruber, C.L. Sabine, C.L. Le Quéré, R.M. Key, and J. Boutin. 2001. Estimates of anthropogenic carbon uptake from four three-dimensional global ocean models. Global Biogeochem. Cy. 15(1):43–60.

Arctic Sea Ice

Identification

1. Indicator Description

This indicator tracks the extent and age of sea ice in the Arctic Ocean. The extent of area covered by Arctic sea ice is considered a particularly sensitive indicator of global climate because a warmer climate will reduce the amount of sea ice present. The proportion of sea ice in each age category can indicate the relative stability of Arctic conditions as well as susceptibility to melting events.

Components of this indicator include:

- Changes in the September average extent of sea ice in the Arctic Ocean since 1979 (Figure 1).

- Changes in the proportion of Arctic sea ice in various age categories at the September weekly minimum since 1983 (Figure 2).

2. Revision History

April 2010:   Indicator of Arctic sea ice extent posted.
December 2011:   Updated with data through 2011; age of ice added.
October 2012:   Updated with data through 2012.
December 2013:   Updated with data through 2013.

Data Sources

3. Data Sources

Figure 1 (extent of sea ice) is based on monthly average sea ice extent data provided by the National Snow and Ice Data Center (NSIDC). NSIDC's data are derived from satellite imagery collected and processed by the National Aeronautics and Space Administration (NASA). NSIDC also provided Figure 2 data (age distribution of sea ice), which are derived from weekly NASA satellite imagery and processed by the team of Maslanik and Tschudi at the University of Colorado, Boulder.

4. Data Availability

Figure 1. September Monthly Average Arctic Sea Ice Extent, 1979–2013

Users can access monthly map images, geographic information system (GIS)-compatible map files, and gridded daily and monthly satellite data, along with corresponding metadata, at: http://nsidc.org/data/seaice_index/archives.html. From this page, users can also download monthly extent and area data. From this page, select "FTP Directory" under the "Monthly Extent and Concentration Images" heading, which will lead to a public FTP site (ftp://sidads.colorado.edu/DATASETS/NOAA/G02135). To obtain the September monthly data that were used in this indicator, select the "Sep" directory, then choose the "...area.txt" file with the data. To see a different version of the graph in Figure 1 (plotting percent anomalies rather than square miles), return to the "Sep" directory and open the "...plot.png" image.

NSIDC's Sea Ice Index documentation page (http://nsidc.org/data/docs/noaa/g02135_seaice_index) describes how to download, read, and interpret the data. It also defines database fields and key terminology. Gridded source data can be found at: http://nsidc.org/data/nsidc-0051.html and: http://nsidc.org/data/nsidc-0081.html.

Figure 2. Age of Arctic Sea Ice at Minimum September Week, 1983–2013

NSIDC published a map version of Figure 2 at: http://nsidc.org/arcticseaicenews/2013/10. EPA obtained the data shown in the figure by contacting NSIDC User Services. The data are processed by Dr. James Maslanik and Dr. Mark Tschudi at the University of Colorado, Boulder, and provided to NSIDC. Earlier versions of this analysis appeared in Maslanik et al. (2011) and Maslanik et al. (2007).

Satellite data used in historical and ongoing monitoring of sea ice age can be found at the following websites:

- Defense Meteorological Satellite Program (DMSP) Scanning Multi Channel Microwave Radiometer (SMMR): http://nsidc.org/data/nsidc-0071.html.
- DMSP Special Sensor Microwave/Imager (SSM/I): http://nsidc.org/data/nsidc-0001.html.
- DMSP Special Sensor Microwave Imager and Sounder (SSMIS): http://nsidc.org/data/nsidc-0001.html.
- NASA Advanced Microwave Scanning Radiometer for the Earth Observing System (AMSR-E): http://nsidc.org/data/amsre.
- Advanced Very High Resolution Radiometer (AVHRR): http://nsidc.org/data/avhrr/data_summaries.html.

Age calculations also depend on wind measurements and on buoy-based measurements and motion vectors. Wind measurements (as surface flux data) are available at: www.esrl.noaa.gov/psd/data/reanalysis/reanalysis.shtml. Data and metadata are available online at: http://iabp.apl.washington.edu/data.html and: http://nsidc.org/data/nsidc-0116.html.

Methodology

5. Data Collection

This indicator is based on maps of sea ice extent in the Arctic Ocean and surrounding waters, which were developed using brightness temperature imagery collected by satellites. Data from October 1978 through June 1987 were collected using the Nimbus-7 SMMR instrument, and data since July 1987 have been collected using a series of successor SSM/I instruments. In 2008, the SSMIS replaced the SSM/I as the source for sea ice products. These instruments can identify the presence of sea ice because sea ice and open water have different passive microwave signatures. The record has been supplemented with data from AMSR-E, which operated from 2003 to 2011.

The satellites that supply data for this indicator orbit the Earth continuously, collecting images that can be used to generate daily maps of sea ice extent. They are able to map the Earth's surface with a resolution of 25 kilometers. The resultant maps have a nominal pixel area of 625 square kilometers. Because of the curved map projection, however, actual pixel sizes range from 382 to 664 square kilometers.

The satellites that collect the data cover most of the Arctic region in their orbital paths. However, the sensors cannot collect data from a circular area immediately surrounding the North Pole due to orbit inclination. From 1978 through June 1987, this "pole hole" measured 1.19 million square kilometers. Since July 1987 it has measured 0.31 million square kilometers. For more information about this spatial gap and how it is corrected in the final data, see Section 6.

To calculate the age of ice (Figure 2), the SSM/I, SMMR, and AMSR-E imagery have been supplemented with three additional data sets:

- AVHRR satellite data, which come from an optical sensing instrument that can measure sea ice temperature and heat flux, which in turn can be used to estimate thickness. AVHRR also covers the "pole hole."

- Maps of wind speed and direction at 10 meters above the Earth's surface, which were compiled by the National Oceanic and Atmospheric Administration's (NOAA's) National Centers for Environmental Prediction (NCEP).

- Motion vectors that trace how parcels of sea ice move, based on data collected by the International Arctic Buoy Programme (IABP). Since 1955, the IABP has deployed a network of 14 to 30 *in situ* buoys in the Arctic Ocean that provide information about movement rates at six-hour intervals.

For documentation of passive microwave satellite data collection methods, see the summary and citations at: http://nsidc.org/data/docs/noaa/g02135_seaice_index. For further information on AVHRR imagery, see: http://noaasis.noaa.gov/NOAASIS/ml/avhrr.html. For motion tracking methods, see Maslanik et al. (2011), Fowler et al. (2004), and: http://nsidc.org/data/nsidc-0116.html.

6. Indicator Derivation

Figure 1. September Monthly Average Arctic Sea Ice Extent, 1979–2013

Satellite data are used to develop daily ice extent and concentration maps using an algorithm developed by NASA. Data are evaluated within grid cells on the map. Image processing includes quality control features such as two weather filters based on brightness temperature ratios to screen out false positives over open water, an ocean mask to eliminate any remaining sea ice in regions where sea ice is not expected, and a coastal filter to eliminate most false positives associated with mixed land/ocean grid cells.

From each daily map, analysts calculate the total "extent" and "area" covered by ice. These terms are defined differently as a result of how they address those portions of the ocean that are partially but not completely frozen:

- **Extent** is the total area covered by all pixels on the map that have at least 15 percent ice concentration, which means at least 15 percent of the ocean surface within that pixel is frozen over. The 15 percent concentration cutoff for extent is based on validation studies that showed that a 15 percent threshold provided the best approximation of the "true" ice edge and the lowest bias. In practice, most of the area covered by sea ice in the Arctic far exceeds the 15 percent threshold, so using a higher cutoff (e.g., 20 or 30 percent) would yield different totals but similar overall trends (for example, see Parkinson et al., 1999).

- **Area** represents the actual surface area covered by ice. If a pixel's area were 600 square kilometers and its ice concentration were 75 percent, then the ice area for that pixel would be 450 square kilometers. At any point in time, total ice area will always be less than total ice extent.

EPA's indicator addresses extent rather than area. Both of these measurements are valid ways to look at trends in sea ice, but in this case, EPA chose to look at the time series for extent because it is more complete than the time series for area. In addition, the available area data set does not include the "pole hole" (the area directly above the North Pole that the satellites cannot cover), and the size of this unmapped region changed as a result of the instrumentation change in 1987, creating a discontinuity in the area data. In contrast, the extent time series assumes that the entire "pole hole" area is covered with at least 15 percent ice, which is a reasonable assumption based on other observations of this area.

NASA's processing algorithm includes steps to deal with occasional days with data gaps due to satellite or sensor outages. These days were removed from the time series and replaced with interpolated values based on the total extent of ice on the surrounding days.

From daily maps and extent totals, NSIDC calculated monthly average extent in square kilometers. EPA converted these values to square miles to make the results accessible to a wider audience. By relying on monthly averages, this indicator smoothes out some of the variability inherent in daily measurements.

Figure 1 shows trends in September average sea ice extent. September is when Arctic sea ice typically reaches its annual minimum, after melting during the summer months. By looking at the month with the smallest extent of sea ice, this indicator focuses attention on the time of year when limiting conditions would most affect wildlife and human societies in the Arctic region.

This indicator does not attempt to estimate values from before the onset of regular satellite mapping in October 1978 (which makes 1979 the first year with September data for this indicator). It also does not attempt to project data into the future.

For documentation of the NASA Team algorithm used to process the data, see Cavalieri et al. (1984) and: http://nsidc.org/data/nsidc-0051.html. For more details about NSIDC methods, see the Sea Ice Index documentation and related citations at: http://nsidc.org/data/docs/noaa/g02135_seaice_index.

Other months of the year were considered for this indicator, but EPA chose to focus on September, which is when the extent of ice reaches its annual minimum. September extent is often used as an indicator. One reason is because as temperatures start to get colder, there may be less meltwater on the surface than during the previous summer months, thus leading to more reliable remote sensing of ice extent. Increased melting during summer months leads to changes in the overall character of the ice (i.e., age and thickness) and these changes have implications throughout the year. Thus, September conditions are particularly important for the overall health of Arctic sea ice.

Evidence shows that the extent of Arctic sea ice has declined in all months of the year. Comiso (2012) examined the seasonal pattern in Arctic sea ice extent for three decadal periods plus the years 2007, 2009, and 2010 and found declines throughout the year. Figure TD-1 shows monthly means based on an analysis from NSIDC—the source of data for this indicator. It reveals that Arctic sea ice extent has generally declined over time in all months, with the most pronounced decline in the summer and fall.

Figure TD-1. Monthly Arctic Sea Ice Extent, 1978/1979–2013

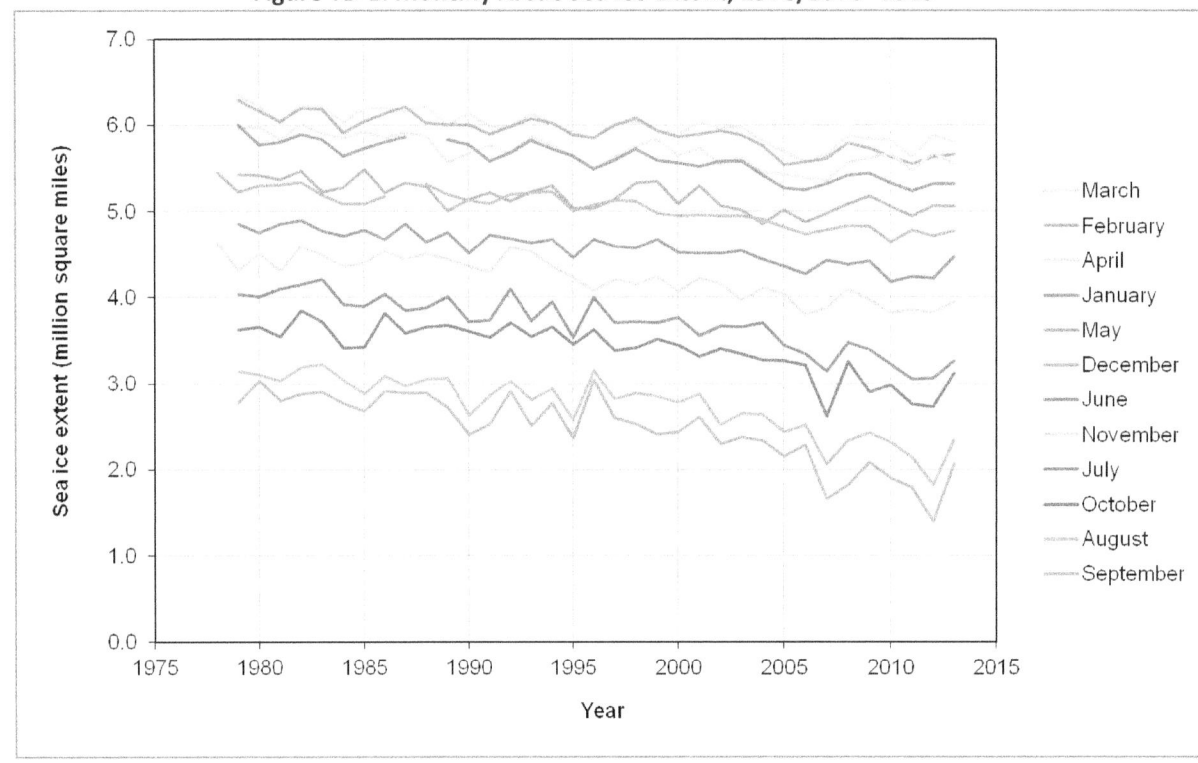

Data source: NSIDC: _http://nsidc.org/data/seaice_index/archives.html_. Accessed January 2014.

Figure 2. Age of Arctic Sea Ice at Minimum September Week, 1983–2013

A research team at the University of Colorado at Boulder processes daily sequential SSM/I, SMMR, AMSR-E, and AVHRR satellite data from NASA, then produces maps using a grid with 12 km-by-12 km cells. The AVHRR data help to fill the "pole hole" and provide information about the temperature and thickness of the ice. Like Figure 1, this method classifies a pixel as "ice" if at least 15 percent of the ocean surface within the area is frozen over. Using buoy data from the IABP, motion vectors for the entire region are blended via optimal interpolation and mapped on the gridded field. NCEP wind data are also incorporated at this stage, with lower weighting during winter and higher weighting during summer, when surface melt limits the performance of the passive microwave data. Daily ice extent and motion vectors are averaged on a weekly basis. Once sea ice reaches its annual minimum extent (typically in early September), the ice is documented as having aged by one year. For further information on data processing methods, see Maslanik et al. (2011), Maslanik et al. (2007), and Fowler et al. (2004). Although the most recently published representative study does not utilize AMSR-E brightness data or

NCEP wind data for the calculation of ice motion, the results presented in Figure 2 and the NSIDC website incorporate these additional sources.

Figure 2 shows the extent of ice that falls into several age categories. Whereas Figure 1 extends back to 1979, Figure 2 can show trends only back to 1983 because it is not possible to know how much ice is five or more years old (the oldest age class shown) until parcels of ice have been tracked for at least five years. Regular satellite data collection did not begin until October 1978, which makes 1983 the first year in which September minimum ice can be assigned to the full set of age classes shown in Figure 2.

7. Quality Assurance and Quality Control

Image processing includes a variety of quality assurance and quality control (QA/QC) procedures, including steps to screen out false positives. These procedures are described in NSIDC's online documentation at: http://nsidc.org/data/docs/noaa/g02135_seaice_index as well as in some of the references cited therein.

NSIDC Arctic sea ice data have three levels of processing for quality control. NSIDC's most recent data come from the Near Real-Time SSM/I Polar Gridded Sea Ice Concentrations (NRTSI) data set. NRTSI data go through a first level of calibration and quality control to produce a preliminary data product. The final data are processed by NASA's Goddard Space Flight Center (GSFC), which uses a similar process but applies a higher level of QC. Switching from NRTSI to GSFC data can result in slight changes in the total extent values—on the order of 50,000 square kilometers or less for total sea ice extent.

Because GSFC processing requires several months' lag time, Figure 1 reports GSFC data for the years 1979 to 2012 and a NRTSI data point for 2013. At the time EPA published this report, the GSFC data for 2013 had not yet been finalized.

Analysis

8. Comparability Over Time and Space

Both figures for this indicator are based on data collection methods and processing algorithms that have been applied consistently over time and space. NASA's satellites cover the entire area of interest with the exception of a "hole" at the North Pole for Figure 1. Even though the size of this hole has changed over time, EPA's indicator uses a data set that corrects for this discontinuity.

The total extent shown in Figure 2 (the sum of all the stacked areas) differs from the total extent in Figure 1 because Figure 2 shows conditions during the specific week in September when minimum extent is reached, while Figure 1 shows average conditions over the entire month of September. It would not make sense to convert Figure 2 to a monthly average for September because all ice is "aged" one year as soon as the minimum has been achieved, which creates a discontinuity after the minimum week.

9. Data Limitations

Factors that may impact the confidence, application, or conclusions drawn from this indicator are as follows:

1. Variations in sea ice are not entirely due to changes in temperature. Other conditions, such as fluctuations in oceanic and atmospheric circulation and typical annual and decadal variability, can also affect the extent of sea ice, and by extension the sea ice age indicator.

2. Changes in the age and thickness of sea ice—for example, a trend toward younger or thinner ice—might increase the rate at which ice melts in the summer, making year-to-year comparisons more complex.

3. Many factors can diminish the accuracy of satellite mapping of sea ice. Although satellite instruments and processing algorithms have improved somewhat over time, applying these new methods to established data sets can lead to trade-offs in terms of reprocessing needs and compatibility of older data. Hence, this indicator does not use the highest-resolution imagery or the newest algorithms. Trends are still accurate, but should be taken as a general representation of trends in sea ice extent, not an exact accounting.

4. As described in Section 6, the threshold used to determine extent—15 percent ice cover within a given pixel—represents an arbitrary cutoff without a particular scientific significance. Nonetheless, studies have found that choosing a different threshold would result in a similar overall trend. Thus, the most important part of Figure 1 is not the absolute extent reported for any given year, but the size and shape of the trend over time.

5. Using ice surface data and motion vectors allows only the determination of a maximum sea ice age. Thus, as presented, the Figure 2 indicator indicates the age distribution of sea ice only on the surface, and is not necessarily representative of the age distribution of the total sea ice volume.

10. Sources of Uncertainty

NSIDC has calculated standard deviations along with each monthly ice concentration average. NSIDC's Sea Ice Index documentation (http://nsidc.org/data/docs/noaa/g02135_seaice_index) describes several analyses that have examined the accuracy and uncertainty of passive microwave imagery and the NASA Team algorithm used to create this indicator. For example, a 1991 analysis estimated that ice concentrations measured by passive microwave imagery are accurate to within 5 to 9 percent, depending on the ice being imaged. Another study suggested that the NASA Team algorithm underestimates ice extent by 4 percent in the winter and more in summer months. A third study that compared the NASA Team algorithm with new higher-resolution data found that the NASA Team algorithm underestimates ice extent by an average of 10 percent. For more details and study citations, see: http://nsidc.org/data/docs/noaa/g02135_seaice_index. Certain types of ice conditions can lead to larger errors, particularly thin or melting ice. For example, a melt pond on an ice floe might be mapped as open water. The instruments also can have difficulty distinguishing the interface between ice and snow or a diffuse boundary between ice and open water. Using the September minimum minimizes many of these effects because melt ponds and the ice surface become largely frozen by then. These errors do not affect trends and relative changes from year to year.

NSIDC has considered using a newer algorithm that would process the data with greater certainty, but doing so would require extensive research and reprocessing, and data from the original instrument (pre-1987) might not be compatible with some of the newer algorithms that have been proposed. Thus, for the time being, this indicator uses the best available science to provide a multi-decadal representation of trends in Arctic sea ice extent. The overall trends shown in this indicator have been corroborated by numerous other sources, and readers should feel confident that the indicator provides an accurate overall depiction of trends in Arctic sea ice over time.

Accuracy of ice motion vectors depends on the error in buoy measurements, wind fields, and satellite images. Given that buoy locational readings are taken every six hours, satellite images are 24-hour averages, and a "centimeters per second" value is interpolated based on these readings, accuracy depends on the error of the initial position and subsequent readings. NSIDC proposes that "the error would be less than 1 cm/sec for the average velocity over 24 hours" (http://nsidc.org/data/docs/daac/nsidc0116_icemotion/buoy.html).

11. Sources of Variability

Many factors contribute to variability in this indicator. In constructing the indicator, several choices have been made to minimize the extent to which this variability affects the results. The apparent extent of sea ice can vary widely from day to day, both due to real variability in ice extent (growth, melting, and movement of ice at the edge of the ice pack) and due to ephemeral effects such as weather, clouds and water vapor, melt on the ice surface, and changes in the character of the snow and ice surface. The intensity of Northern Annular Mode (NAM) conditions and changes to the Arctic Oscillation also have a strong year-to-year impact on ice movement. Under certain conditions, older ice might move to warmer areas and be subject to increased melting. Weather patterns can also affect the sweeping of icebergs out of the Arctic entirely. For a more complete description of major thermodynamic processes that impact ice longevity, see Maslanik et al. (2007) and Rigor and Wallace (2004).

According to NSIDC's documentation at: http://nsidc.org/data/docs/noaa/g02135_seaice_index, extent is a more reliable variable than ice concentration or area. The weather and surface effects described above can substantially impact estimates of ice concentration, particularly near the edge of the ice pack. Extent is a more stable variable because it simply registers the presence of at least a certain percentage of sea ice in a grid cell (15 percent). For example, if a particular pixel has an ice concentration of 50 percent, outside factors could cause the satellite to measure the concentration very differently, but as long as the result is still greater than the percent threshold, this pixel will be correctly accounted for in the total "extent." Monthly averages also help to reduce some of the day-to-day "noise" inherent in sea ice measurements.

12. Statistical/Trend Analysis

This indicator does not report on the slope of the apparent trends in September sea ice extent and age distribution, nor does it calculate the statistical significance of these trends.

References

Cavalieri, D.J., P. Gloersen, and W.J. Campbell. 1984. Determination of sea ice parameters with the NIMBUS-7 SMMR. J. Geophys. Res. 89(D4):5355–5369.

Comiso, J. 2012. Large decadal decline of the Arctic multiyear ice cover. J. Climate 25(4):1176–1193.

Fowler, C., W.J. Emery, and J. Maslanik. 2004. Satellite-derived evolution of Arctic sea ice age: October 1978 to March 2003. IEEE Geosci. Remote S. 1(2):71–74.

Maslanik, J.A., C. Fowler, J. Stroeve, S. Drobot, J. Zwally, D. Yi, and W. Emery. 2007. A younger, thinner Arctic ice cover: Increased potential for rapid, extensive sea-ice loss. Geophys. Res. Lett. 34:L24501.

Maslanik, J., J. Stroeve, C. Fowler, and W. Emery. 2011. Distribution and trends in Arctic sea ice age through spring 2011. Geophys. Res. Lett. 38:L13502.

Parkinson, C.L., D.J. Cavalieri, P. Gloersen, H.J. Zwally, and J.C. Comiso. 1999. Arctic sea ice extents, areas, and trends, 1978–1996. J. Geophys. Res. 104(C9):20,837–20,856.

Rigor, I.G., and J.M. Wallace. 2004. Variations in the age of Arctic sea-ice and summer sea-ice extent. Geophys. Res. Lett. 31:L09401. http://iabp.apl.washington.edu/research_seaiceageextent.html.

Glaciers

Identification

1. Indicator Description

This indicator examines the balance between snow accumulation and melting in glaciers, and describes how the size of glaciers around the world has changed since 1945. On a local and regional scale, changes in glaciers have implications for ecosystems and people who depend on glacier-fed streamflow. On a global scale, loss of ice from glaciers contributes to sea level rise. Glaciers are important as an indicator of climate change because physical changes in glaciers—whether they are growing or shrinking, advancing or receding—provide visible evidence of changes in temperature and precipitation.

Components of this indicator include:

- Cumulative trends in the mass balance of reference glaciers worldwide over the past 65 years (Figure 1).
- Cumulative trends in the mass balance of three U.S. glaciers over the past half-century (Figure 2).

2. Revision History

April 2010:	Indicator posted.
December 2011:	Updated with data through 2010.
April 2012:	Replaced Figure 1 with data from a new source: the World Glacier Monitoring Service.
June 2012:	Updated Figure 2 with data through 2010 for South Cascade Glacier.
May 2014:	Updated Figures 1 and 2 with data through 2012.

Data Sources

3. Data Sources

Figure 1 shows the average cumulative mass balance of a global set of reference glaciers, which was originally published by the World Glacier Monitoring Service (WGMS) (2013). Measurements were collected by a variety of academic and government programs and compiled by WGMS.

The U.S. Geological Survey (USGS) Benchmark Glacier Program provided the data for Figure 2, which shows the cumulative mass balance of three U.S. "benchmark" glaciers where long-term monitoring has taken place.

4. Data Availability

Figure 1. Average Cumulative Mass Balance of "Reference" Glaciers Worldwide, 1945–2012

A version of Figure 1 with data through 2011 was published in WGMS (2013). Preliminary values for 2012 were posted by WGMS at: www.wgms.ch/mbb/sum12.html. Some recent years are associated with a reduced number of associated reference glaciers (e.g., 34 instead of the full set of 37). EPA obtained the data in spreadsheet form from the staff of WGMS, which can be contacted via their website: www.wgms.ch/access.html.

Raw measurements of glacier surface parameters around the world have been recorded in a variety of formats. Some data are available in online databases such as the World Glacier Inventory (http://nsidc.org/data/glacier_inventory/index.html). Some raw data are also available in studies by USGS. WGMS maintains perhaps the most comprehensive record of international observations. Some of these observations are available in hard copy only; others are available through an online data browser at: www.wgms.ch/metadatabrowser.html.

Figure 2. Cumulative Mass Balance of Three U.S. Glaciers, 1958–2012

A cumulative net mass balance data set is available on the USGS benchmark glacier website at: http://ak.water.usgs.gov/glaciology/all_bmg/3glacier_balance.htm. Because the online data are not necessarily updated every time a correction or recalculation is made, EPA obtained the most up-to-date data for Figure 2 directly from USGS. More detailed metadata and measurements from the three benchmark glaciers can be found on the USGS website at: http://ak.water.usgs.gov/glaciology.

Methodology

5. Data Collection

This indicator provides information on the cumulative change in mass balance of numerous glaciers over time. Glacier mass balance data are calculated based on a variety of measurements at the surface of a glacier, including measurements of snow depths and snow density. These measurements help glaciologists determine changes in snow and ice accumulation and ablation that result from snow precipitation, snow compaction, freezing of water, melting of snow and ice, calving (i.e., ice breaking off from the tongue or leading edge of the glacier), wind erosion of snow, and sublimation from ice (Mayo et al., 2004). Both surface size and density of glaciers are measured to produce net mass balance data. These data are reported in meters of water equivalent (mwe), which corresponds to the average change in thickness over the entire surface area of the glacier. Because snow and ice can vary in density (depending on the degree of compaction, for example), converting to the equivalent amount of liquid water provides a more consistent metric.

Measurement techniques have been described and analyzed in many peer-reviewed studies, including Josberger et al. (2007). Most long-term glacier observation programs began as part of the International Geophysical Year in 1957–1958.

Figure 1. Average Cumulative Mass Balance of "Reference" Glaciers Worldwide, 1945–2012

The global trend is based on data collected at 37 reference glaciers around the world, which are identified in Table TD-1.

Table TD-1. Reference Glaciers Included in Figure 1

Continent	Region	Glaciers
North America	Alaska	Gulkana, Wolverine
North America	Pacific Coast Ranges	Place, South Cascade, Helm, Lemon Creek, Peyto
North America	Canadian High Arctic	Devon Ice Cap NW, Meighen Ice Cap, White
South America	Andes	Echaurren Norte
Europe	Svalbard	Austre Broeggerbreen, Midtre Lovénbreen
Europe	Scandinavia	Engabreen, Alfotbreen, Nigardsbreen, Grasubreen, Storbreen, Hellstugubreen, Hardangerjoekulen, Storglaciaeren
Europe	Alps	Saint Sorlin, Sarennes, Argentière, Silvretta, Gries, Stubacher Sonnblickkees, Vernagtferner, Kesselwandferner, Hintereisferner, Caresèr
Europe/Asia	Caucasus	Djankuat
Asia	Altai	No. 125 (Vodopadniy), Maliy Aktru, Leviy Aktru
Asia	Tien Shan	Ts. Tuyuksuyskiy, Urumqi Glacier No.1

WGMS chose these 37 reference glaciers because they all had at least 30 years of continuous mass balance records (WGMS, 2013). As the small graph at the bottom of Figure 1 shows, some of these glaciers have data extending as far back as the 1940s. WGMS did not include data from glaciers that are dominated by non-climatic factors, such as surge dynamics or calving. Because of data availability and the distribution of glaciers worldwide, WGMS's compilation is dominated by the Northern Hemisphere.

All of the mass balance data that WGMS compiled for this indicator are based on the direct glaciological method (Østrem and Brugman, 1991), which involves manual measurements with stakes and pits at specific points on each glacier's surface.

Figure 2. Cumulative Mass Balance of Three U.S. Glaciers, 1958–2012

Figure 2 shows data collected at the three glaciers studied by USGS's Benchmark Glacier Program. All three glaciers have been monitored for many decades. USGS chose them because they represent typical glaciers found in their respective regions: South Cascade Glacier in the Pacific Northwest (a continental glacier), Wolverine Glacier in coastal Alaska (a maritime glacier), and Gulkana Glacier in inland Alaska (a continental glacier). Hodge et al. (1998) and Josberger et al. (2007) provide more information about the locations of these glaciers and why USGS selected them for the benchmark monitoring program.

USGS collected repeated measurements at each of the glaciers to determine the various parameters that can be used to calculate cumulative mass balance. Specific information on sampling design at each of the three glaciers is available in Bidlake et al. (2010) and Van Beusekom et al. (2010). Measurements are collected at specific points on the glacier surface, designated by stakes.

Data for South Cascade Glacier are available beginning in 1959 (relative to conditions in 1958) and for Gulkana and Wolverine Glaciers beginning in 1966 (relative to conditions in 1965). Glacier monitoring methodology has evolved over time based on scientific reanalysis, and cumulative net mass balance data for these three glaciers are routinely updated as glacier measurement methodologies improve and more information becomes available. Several papers that document data updates through time are available on the USGS benchmark glacier website at: http://ak.water.usgs.gov/glaciology.

6. Indicator Derivation

For this indicator, glacier surface measurements have been used to determine the net change in mass balance from one year to the next, referenced to the previous year's summer surface measurements. The indicator documents changes in mass and volume rather than total mass or volume of each glacier because the latter is more difficult to determine accurately. Thus, the indicator is not able to show how the magnitude of mass balance change relates to the overall mass of the glacier (e.g., what percentage of the glacier's mass has been lost).

Glaciologists convert surface measurements to mass balance by interpolating measurements over the glacier surface geometry. Two different interpolation methods can be used: conventional balance and reference-surface balance. In the conventional balance method, measurements are made at the glacier each year to determine glacier surface geometry, and other measurements are interpolated over the annually modified geometry. The reference-surface balance method does not require that glacier geometry be redetermined each year. Rather, glacier surface geometry is determined once, generally the first year that monitoring begins, and the same geometry is used each of the following years. A more complete description of conventional balance and reference-surface balance methods is given in Harrison et al. (2009).

Mass balance is typically calculated over a balance year, which begins at the onset of snow and ice accumulation. For example, the balance year at Gulkana Glacier starts and ends in September of each year. Thus, the balance year beginning in September 2011 and ending in September 2012 is called "balance year 2012." Annual mass balance changes are confirmed based on measurements taken the following spring.

Figure 1. Average Cumulative Mass Balance of "Reference" Glaciers Worldwide, 1945–2012

The graph shows the average cumulative mass balance of WGMS's reference glaciers over time. The number of reference glaciers included in this calculation varies by year, but it is still possible to generate a reliable time series because the figure shows an average across all of the glaciers measured, rather than a sum. No attempt was made to extrapolate from the observed data in order to calculate a cumulative global change in mass balance.

Figure 2. Cumulative Mass Balance of Three U.S. Glaciers, 1958–2012

At each of the three benchmark glaciers, changes in mass balance have been summed over time to determine the cumulative change in mass balance since a reference year. For the sake of comparison, all three glaciers use a reference year of 1965, which is set to zero. Thus, a negative value in a later year means the glacier has lost mass since 1965. All three time series in Figure 2 reflect the conventional mass balance method, as opposed to the reference-surface method. No attempt has been made to project the results for the three benchmark glaciers to other locations. See Bidlake et al. (2010), Van Beusekom et al. (2010), and sources cited therein for further description of analytical methods.

In the past, USGS formally designated annual mass balance estimates as preliminary or final. USGS no longer does this, choosing instead to continually refine and update mass balance estimates according to the best available science and data. Accordingly, USGS provides new data to support regular updates of this indicator with measurements that are comparable across glaciers. USGS is currently consolidating glacier records to better harmonize calculation methods across space and time. Future updates of EPA's indicator will reflect this harmonization.

7. Quality Assurance and Quality Control

The underlying measurements for Figure 1 come from a variety of data collection programs, each with its own procedures for quality assurance and quality control (QA/QC). WGMS also has its own requirements for data quality. For example, WGMS incorporates only measurements that reflect the direct glaciological method (Østrem and Brugman, 1991).

USGS periodically reviews and updates the mass balance data shown in Figure 2. For example, in Fountain et al. (1997), the authors explain that mass balance should be periodically compared with changes in ice volume, as the calculations of mass balance are based on interpolation of point measurements that are subject to error. In addition, March (2003) describes steps that USGS takes to check the weighting of certain mass balance values. This weighting allows USGS to convert point values into glacier-averaged mass balance values.

Ongoing reanalysis of glacier monitoring methods, described in several of the reports listed on USGS's website (http://ak.water.usgs.gov/glaciology), provides an additional level of quality control for data collection.

Analysis

8. Comparability Over Time and Space

Glacier monitoring methodology has evolved over time based on scientific reanalysis of methodology. Peer-reviewed studies describing the evolution of glacier monitoring are listed in Mayo et al. (2004). Figure 2 accounts for these changes, as USGS periodically reanalyzes past data points using improved methods.

The reference glaciers tracked in Figure 1 reflect a variety of methods over time and space, and it is impractical to adjust for all of these small differences. However, as a general indication of trends in glacier mass balance, Figure 1 shows a clear pattern whose strength is not diminished by the inevitable variety of underlying sources.

9. Data Limitations

Factors that may impact the confidence, application, or conclusions drawn from this indicator are as follows:

1. Slightly different methods of measurement and interpolation have been used at different glaciers, making direct year-to-year comparisons of change in cumulative net mass balance or volume difficult. Overall trends among glaciers can be compared, however.

2. The number of glaciers with data available to calculate mass balance in Figure 1 decreases as one goes back in time. Thus, averages from the 1940s to the mid-1970s rely on a smaller set of reference glaciers than the full 37 compiled in later years.

3. The relationship between climate change and glacier mass balance is complex, and the observed changes at a specific glacier might reflect a combination of global and local climate variations.

4. Records are available from numerous other individual glaciers in the United States, but many of these other records lack the detail, consistency, or length of record provided by the USGS benchmark glaciers program. USGS has collected data on these three glaciers for decades using consistent methods, and USGS experts suggest that at least a 30-year record is necessary to provide meaningful statistics. Due to the complicated nature of glacier behavior, it is difficult to assess the significance of observed trends over shorter periods (Josberger et al., 2007).

10. Sources of Uncertainty

Glacier measurements have inherent uncertainties. For example, maintaining a continuous and consistent data record is difficult because the stakes that denote measurement locations are often distorted by glacier movement and snow and wind loading. Additionally, travel to measurement sites is dangerous and inclement weather can prevent data collection during the appropriate time frame. In a cumulative time series, such as the analyses presented in this indicator, the size of the margin of error grows with time because each year's value depends on all of the preceding years.

Figure 1. Average Cumulative Mass Balance of "Reference" Glaciers Worldwide, 1945–2012

Uncertainties have been quantified for some glacier mass balance measurements, but not for the combined time series shown in Figure 1. WGMS (2013) has identified greater quantification of uncertainty in mass balance measurements as a key goal for future research.

Figure 2. Cumulative Mass Balance of Three U.S. Glaciers, 1958–2012

Annual mass balance measurements for the three USGS benchmark glaciers usually have an estimated error of ±0.1 to ±0.2 meters of water equivalent (Josberger et al., 2007). Error bars for the two Alaskan glaciers are plotted in Van Beusekom et al. (2010). Further information on error estimates is given in Bidlake et al. (2010) and Van Beusekom et al. (2010). Harrison et al. (2009) describe error estimates related to interpolation methods.

11. Sources of Variability

Glacier mass balance can reflect year-to-year variations in temperature, precipitation, and other factors. Figure 2 shows some of this year-to-year variability, while Figure 1 shows less variability because the change in mass balance has been averaged over many glaciers around the world. In both cases, the availability of several decades of data allows the indicator to show long-term trends that exceed the "noise" produced by interannual variability. In addition, the period of record is longer than the period of key multi-year climate oscillations such as the Pacific Decadal Oscillation and El Niño–Southern Oscillation, meaning the trends shown in Figures 1 and 2 are not simply the product of decadal-scale climate oscillations.

12. Statistical/Trend Analysis

Figures 1 and 2 both show a cumulative loss of mass or volume over time, from which analysts can derive an average annual rate of change. Confidence bounds are not provided for the trends in either figure, although both Bidlake et al. (2010) and Van Beusekom et al. (2010) cite clear evidence of a decline in mass balance at U.S. benchmark glaciers over time.

References

Bidlake, W.R., E.G. Josberger, and M.E. Savoca. 2010. Modeled and measured glacier change and related glaciological, hydrological, and meteorological conditions at South Cascade Glacier, Washington, balance and water years 2006 and 2007. U.S. Geological Survey Scientific Investigations Report 2010–5143. http://pubs.usgs.gov/sir/2010/5143.

Fountain, A.G., R.M. Krimmel, and D.C. Trabant. 1997. A strategy for monitoring glaciers. U.S. Geological Survey Circular 1132.

Harrison, W.D., L.H. Cox, R. Hock, R.S. March, and E.C. Petit. 2009. Implications for the dynamic health of a glacier from comparison of conventional and reference-surface balances. Ann. Glaciol. 50:25–30.

Hodge, S.M., D.C. Trabant, R.M. Krimmel, T.A. Heinrichs, R.S. March, and E.G. Josberger. 1998. Climate variations and changes in mass of three glaciers in western North America. J. Climate 11:2161–2217.

Josberger, E.G., W.R. Bidlake, R.S. March, and B.W. Kennedy. 2007. Glacier mass-balance fluctuations in the Pacific Northwest and Alaska, USA. Ann. Glaciol. 46:291–296.

March, R.S. 2003. Mass balance, meteorology, area altitude distribution, glacier-surface altitude, ice motion, terminus position, and runoff at Gulkana Glacier, Alaska, 1996 balance year. U.S. Geological Survey Water-Resources Investigations Report 03-4095.

Mayo, L.R., D.C. Trabant, and R.S. March. 2004. A 30-year record of surface mass balance (1966–95) and motion and surface altitude (1975–95) at Wolverine Glacier, Alaska. U.S. Geological Survey Open-File Report 2004-1069.

Østrem, G., and M. Brugman. 1991. Glacier mass-balance measurements: A manual for field and office work. National Hydrology Research Institute (NHRI), NHRI Science Report No. 4.

Van Beusekom, A.E., S.R. O'Neel, R.S. March, L.C. Sass, and L.H. Cox. 2010. Re-analysis of Alaskan benchmark glacier mass-balance data using the index method. U.S. Geological Survey Scientific Investigations Report 2010–5247. http://pubs.usgs.gov/sir/2010/5247.

WGMS (World Glacier Monitoring Service). 2013. Glacier mass balance bulletin no. 12 (2010–2011). Zemp, M., S.U. Nussbaumer, K. Naegeli, I. Gärtner-Roer, F. Paul, M. Hoelzle, and W. Haeberli (eds.). ICSU(WDS)/IUGG(IACS)/UNEP/UNESCO/WMO. Zurich, Switzerland: World Glacier Monitoring Service. www.wgms.ch/mbb/mbb12/wgms_2013_gmbb12.pdf.

Lake Ice

Identification

1. Indicator Description

This indicator tracks when selected lakes in the United States froze and thawed each year between approximately 1850 and 2012. The formation of ice cover on lakes in the winter and its disappearance the following spring depends on climate factors such as air temperature, cloud cover, and wind. Conditions such as heavy rains or snowmelt in locations upstream or elsewhere in the watershed also affect the length of time a lake is frozen. Thus, ice formation and breakup dates are relevant indicators of climate change. If lakes remain frozen for longer periods, it can signify that the climate is cooling. Conversely, shorter periods of ice cover suggest a warming climate.

Components of this indicator include:

- First freeze dates of selected U.S. lakes since 1850 (Figure 1).
- Ice breakup dates of selected U.S. lakes since 1850 (Figure 2).
- Trends in ice breakup dates of selected U.S. lakes since 1905 (Figure 3).

2. Revision History

April 2010:   Indicator posted.
December 2013:   Updated with data through winter 2012–2013, added seven lakes, removed one
                   lake due to discontinued data (Lake Michigan at Traverse City), removed original
                   Figure 1 (duration), and added new Figure 3 (map showing long-term rates of
                   change in thaw dates).

Data Sources

3. Data Sources

This indicator is mainly based on data from the Global Lake and River Ice Phenology Database, which was compiled by the North Temperate Lakes Long Term Ecological Research program at the Center for Limnology at the University of Wisconsin–Madison from data submitted by participants in the Lake Ice Analysis Group (LIAG). The database is hosted on the Web by the National Snow and Ice Data Center (NSIDC), and it currently contains ice cover data for 750 lakes and rivers throughout the world, some with records as long as 150 years.

Data for many of the selected lakes have not been submitted to the Global Lake and River Ice Phenology Database since 2005. Thus, the most recent data points were obtained from the organizations that originally collected or compiled the observations.

4. Data Availability

Most of the lake ice observations used for this indicator are publicly available from the sources listed below. All of the years listed below and elsewhere in this indicator are presented as base years. Base year 2004 (for example) refers to the winter that begins in 2004, even though the freeze date sometimes occurs in the following year (2005) and the thaw date always occurs in the following year.

The NSIDC's Global Lake and River Ice Phenology Database provides data through 2004 for most lakes, through 2012 for Lake Superior at Bayfield, and through 2010 for Lakes Mendota and Monona. Users can access the NSIDC database at: http://nsidc.org/data/lake_river_ice. Database documentation can be found at: http://nsidc.org/data/docs/noaa/g01377_lake_river_ice.

Users can also view descriptive information about each lake or river in the Global Lake and River Ice Phenology Database. This database contains the following fields, although many records are incomplete:

- Lake or river name
- Lake or river code
- Whether it is a lake or a river
- Continent
- Country
- State
- Latitude (decimal degrees)
- Longitude (decimal degrees)
- Elevation (meters)
- Mean depth (meters)
- Maximum depth (meters)
- Median depth (meters)
- Surface area (square kilometers)
- Shoreline length (kilometers)
- Largest city population
- Power plant discharge (yes or no)
- Area drained (square kilometers)
- Land use code (urban, agriculture, forest, grassland, other)
- Conductivity (microsiemens per centimeter)
- Secchi depth (Secchi disk depth in meters)
- Contributor

Access to the Global Lake and River Ice Phenology Database is unrestricted, but users are encouraged to register so they can receive notification of changes to the database in the future.

Data for years beyond those included in the Global Lake and River Ice Phenology Database come from the following sources:

- Cobbosseecontee Lake, Damariscotta Lake, Moosehead Lake, and Sebago Lake: U.S. Geological Survey (USGS). Data through 2008 come from Hodgkins (2010). Post-2008 data were provided by USGS staff.

- Detroit Lake and Lake Osakis: Minnesota Department of Natural Resources at: www.dnr.state.mn.us/ice_out.

- Geneva Lake: Geneva Lake Environmental Agency Newsletters at: www.genevaonline.com/~glea/newsletters.php.

- Lake George: published by the Lake George Association and collected by the Darrin Freshwater Institute. These data are available online at: www.lakegeorgeassociation.org/who-we-are/documents/IceInOutdatesLakeGeorge2011.pdf.

- Lake Mendota and Lake Monona: North Temperate Lakes Long Term Ecological Research site at: http://lter.limnology.wisc.edu/lakeinfo/ice-data?lakeid=ME and: http://lter.limnology.wisc.edu/lakeinfo/ice-data?lakeid=MO.

- Mirror Lake: available from the Lake Placid Ice Out Benefit contest. The winning dates are published in the Adirondack Daily Enterprise Newspaper. In addition, the Mirror Lake Watershed Association is developing a page to house these data at: www.mirrorlake.net/news-events/ice-in-ice-out-records.

- Otsego Lake: available in the Annual Reports from the State University of New York (SUNY) Oneonta Biological Field Station at: www.oneonta.edu/academics/biofld/publications.asp.

- Shell Lake: provided by Washburn County Clerk.

Methodology

5. Data Collection

This indicator examines two parameters related to ice cover on lakes:

- The annual "ice-on" or freeze date, defined as the first date on which the water body was observed to be completely covered by ice.
- The annual "ice-off," "ice-out," thaw, or breakup date, defined as the date of the last breakup observed before the summer open water phase.

Observers have gathered data on lake ice throughout the United States for many years—in some cases, more than 150 years. The types of observers can vary from one location to another. For example, some observations might have been gathered and published by a local newspaper editor; others compiled by a local resident. Some lakes have benefited from multiple observers, such as residents on both sides of the lake who can compare notes to determine when the lake is completely frozen or thawed. At some locations, observers have kept records of both parameters of interest ("ice-on" and "ice-off"); others might have tracked only one of these parameters.

To ensure sound spatial and temporal coverage, EPA limited this indicator to U.S. water bodies with the longest and most complete historical records. After downloading data for all lakes and rivers within the United States, EPA sorted the data and analyzed each water body to determine data availability for the two parameters of interest. As a result of this analysis, EPA identified 14 water bodies—all lakes—with particularly long and rich records. Special emphasis was placed on identifying water bodies with many consecutive years of data, which can support moving averages and other trend analyses. EPA selected the following 14 lakes for trend analysis:

- Cobbosseecontee Lake, Maine
- Damariscotta Lake, Maine
- Detroit Lake, Minnesota
- Geneva Lake, Wisconsin
- Lake George, New York
- Lake Mendota, Wisconsin
- Lake Monona, Wisconsin
- Lake Osakis, Minnesota
- Lake Superior at Bayfield, Wisconsin
- Mirror Lake, New York
- Moosehead Lake, Maine
- Otsego Lake, New York
- Sebago Lake, Maine
- Shell Lake, Wisconsin

Together, these lakes span parts of the Upper Midwest and the Northeast. The four Maine lakes and Lake Osakis have data for only ice-off, not ice-on, so they do not appear in Figure 1 (first freeze date).

6. Indicator Derivation

Figures 1 and 2. Dates of First Freeze and Ice Thaw for Selected U.S. Lakes, 1850–2012

To smooth out some of the variability in the annual data and to make it easier to see long-term trends in the display, EPA did not plot annual time series but instead calculated nine-year moving averages (arithmetic means) for each of the parameters. EPA chose a nine-year period because it is consistent with other indicators and comparable to the 10-year moving averages used in a similar analysis by Magnuson et al. (2000). Average values are plotted at the center of each nine-year window. For example, the average from 1990 to 1998 is plotted at year 1994. EPA did calculate averages over periods that were missing a few data points. Early years sometimes had sparse data, and the earliest averages were calculated only around the time when many consecutive records started to appear in the record for a given lake.

EPA used endpoint padding to extend the nine-year smoothed lines all the way to the ends of the analysis period for each lake. For example, if annual data were available through 2012, EPA calculated smoothed values centered at 2009, 2010, 2011, and 2012 by inserting the 2008–2012 average into the equation in place of the as-yet-unreported annual data points for 2013 and beyond. EPA used an equivalent approach at the beginning of each time series.

As discussed in Section 4, all data points in Figures 1 and 2 are plotted at the base year, which is the year the winter season began. For the winter of 2010 to 2011, the base year would be 2010, even if a particular lake did not freeze until early 2011.

EPA did not interpolate missing data points. This indicator also does not attempt to portray data beyond the time periods of observation—other than the endpoint padding for the 9-year moving averages—or extrapolate beyond the specific lakes that were selected for the analysis.

Magnuson et al. (2000) and Jensen et al. (2007) describe methods of processing lake ice observations for use in calculating long-term trends.

Figure 3. Change in Ice Thaw Dates for Selected U.S. Lakes, 1905–2012

Long-term trends in ice-off (thaw date) over time were calculated using the Sen slope method as described in Hodgkins (2013). For this calculation, years in which a lake did not freeze were given a thaw date one day earlier than the earliest on record, to avoid biasing the trend by treating the year as missing data. Five lakes had years in which they did not freeze: Geneva, George, Otsego, Sebago, and Superior. Figure 3 shows the total change, which was found by multiplying the slope of the trend line by the total number of years in the period of record.

EPA chose to focus this map on thaw dates, not freeze dates, because several of the target lakes have data for only ice-off, not ice-on. EPA started the Sen slope analysis at 1905 to achieve maximum coverage over a consistent period of record. Choosing an earlier start date would have limited the map to a smaller number of lakes, as several lakes do not have data prior to 1905.

Indicator Development

The version of this indicator that appeared in EPA's *Climate Change Indicators in the United States, 2012* covered eight lakes, and it presented an additional graph that showed the duration of ice cover at the same set of lakes. For the 2014 edition, EPA enhanced this indicator by adding data for seven additional lakes and adding a map with a more rigorous analysis of trends over time. To make room for the map, EPA removed the duration graph, as it essentially just showed the difference between the freeze and thaw dates, which are already shown in other graphs. In fact, in many cases, the data providers determined the duration of ice cover by simply subtracting the freeze date from the thaw date, regardless of whether the lake might have thawed and refrozen during the interim. EPA also removed one lake from the indicator because data are no longer routinely collected there.

7. Quality Assurance and Quality Control

The LIAG performed some basic quality control checks on data that were contributed to the database, making corrections in some cases. Additional corrections continue to be made as a result of user comments. For a description of some recent corrections, see the database documentation at: http://nsidc.org/data/docs/noaa/g01377_lake_river_ice.

Ice observations rely on human judgment. Definitions of "ice-on" and "ice-off" vary, and the definitions used by any given observer are not necessarily documented alongside the corresponding data. Where possible, the scientists who developed the database have attempted to use sources that appear to be consistent from year to year, such as a local resident with a long observation record.

Analysis

8. Comparability Over Time and Space

Historical observations have not been made systematically or according to a standard protocol. Rather, the Global Lake and River Ice Phenology Database—the main source of data for this indicator—represents a systematic effort to compile data from a variety of original sources.

Both parameters were determined by human observations that incorporate some degree of personal judgment. Definitions of these parameters can also vary over time and from one location to another. Human observations provide an advantage, however, in that they enable trend analysis over a much longer time period than can be afforded by more modern techniques such as satellite imagery. Overall, human observations provide the best available record of seasonal ice formation and breakup, and the breadth of available data allows analysis of broad spatial patterns as well as long-term temporal patterns.

9. Data Limitations

Factors that may impact the confidence, application, or conclusions drawn from this indicator are as follows:

1. Although the Global Lake and River Ice Phenology Database provides a lengthy historical record of freeze and thaw dates for a much larger set of lakes and rivers, some records are incomplete, ranging from brief lapses to large gaps in data. Thus, this indicator is limited to 14 lakes with relatively complete historical records. Geographic coverage is limited to sites in four states (Minnesota, Wisconsin, New York, and Maine).

2. Data used in this indicator are all based on visual observations. Records based on visual observations by individuals are open to some interpretation and can reflect different definitions and methods.

3. Historical observations for lakes have typically been made from the shore, which might not be representative of lakes as a whole or comparable to satellite-based observations.

10. Sources of Uncertainty

Ice observations rely on human judgment, and definitions of "ice-on" and "ice-off" vary, which could lead to some uncertainty in the data. For example, some observers might consider a lake to have thawed once they can no longer walk on it, while others might wait until the ice has entirely melted. Observations also depend on one's vantage point along the lake, particularly a larger lake—for example, if some parts of the lake have thawed while others remain frozen. In addition, the definitions used by any given observer are not necessarily documented alongside the corresponding data. Therefore, it is not possible to ensure that all variables have been measured consistently from one lake to another—or even at a single lake over time—and it is also not possible to quantify the true level of uncertainty or correct for such inconsistencies.

Accordingly, the Global Lake and River Ice Phenology Database does not provide error estimates for historical ice observations. Where possible, however, the scientists who developed the database have attempted to use sources that appear to be consistent from year to year, such as a local resident who collects data over a long period. Overall, the Global Lake and River Ice Phenology Database represents the best available data set for lake ice observations, and limiting the indicator to 14 lakes with the most lengthy and complete records should lead to results in which users can have confidence. Consistent patterns of change over time for multiple lakes also provide confidence in the lake ice data.

11. Sources of Variability

For a general idea of the variability inherent in these types of time series, see Magnuson et al. (2000) and Jensen et al. (2007)—two papers that discuss variability and statistical significance for a broader set of lakes and rivers, including some of the lakes in this indicator. Magnuson et al. (2005) discuss variability between lakes, considering the extent to which observed variability reflects factors such as climate patterns, lake morphometry (shape), and lake trophic status. The timing of freeze-up and break-up of ice appears to be more sensitive to air temperature changes at lower latitudes (Livingstone et al., 2010), but despite this, lakes at higher latitudes appear to be experiencing the most rapid reductions in duration of ice cover (Latifovic and Pouliot, 2007).

To smooth out some of the interannual variability and to make it easier to see long-term trends in the display, EPA did not plot annual time series but instead calculated nine-year moving averages (arithmetic means) for each of the parameters, following an approach recommended by Magnuson et al. (2000).

12. Statistical/Trend Analysis

Figure 1 shows data for the nine individual lakes with freeze date data. Figures 2 and 3 show data for all 14 individual lakes. No attempt was made to aggregate the data for multiple lakes. EPA calculated freeze trends over time by ordinary least-squares regression, a common statistical method, to support some of the statements in the "Key Points" section of the indicator. EPA has not calculated the statistical significance of these particular long-term trends, although Magnuson et al. (2000) and Jensen et al. (2007) found that long-term trends in freeze and breakup dates for many lakes were statistically significant ($p < 0.05$). EPA calculated 1905–2012 trends in thaw dates (Figure 3) by computing the Sen slope, an approach used by Hodgkins (2013) and others. Sen slope results were as follows:

- Seven lakes (Cobbosseecontee, Damariscotta, Mirror, Monona, Moosehead, Sebago, and Superior) have trends toward earlier thaw that are significant to a 95 percent level (Mann-Kendall p-value < 0.05).

- Five lakes (Geneva, George, Mendota, Otsego, and Shell) have trends toward earlier thaw that are not significant to a 95 percent level (Mann-Kendall p-value > 0.05), although Lake George's trend is significant to a 90 percent level.

- Two lakes (Detroit and Osakis) have no discernible long-term trend.

A more detailed analysis of trends would potentially consider issues such as serial correlation and short- and long-term persistence.

References

Hodgkins, G.A. 2010. Historical ice-out dates for 29 lakes in New England, 1807–2008. U.S. Geological Survey Open-File Report 2010-1214.

Hodgkins, G.A. 2013. The importance of record length in estimating the magnitude of climatic changes: an example using 175 years of lake ice-out dates in New England. Climatic Change 119:705–718.

Jensen, O.P., B.J. Benson, and J.J. Magnuson. 2007. Spatial analysis of ice phenology trends across the Laurentian Great Lakes region during a recent warming period. Limnol. Oceanogr. 52(5):2013–2026.

Latifovic, R., and D. Pouliot. 2007. Analysis of climate change impacts on lake ice phenology in Canada using the historical satellite data record. Remote Sens. Environ. 106:492–507.

Livingstone, D.M., R. Adrian, T. Blencker, G. George, and G.A. Weyhenmeyer. 2010. Lake ice phenology. In: George, D.G. (ed.). The impact of climate change on European lakes. Aquatic Ecology Series 4:51–62.

Magnuson, J., D. Robertson, B. Benson, R. Wynne, D. Livingstone, T. Arai, R. Assel, R. Barry, V. Card, E. Kuusisto, N. Granin, T. Prowse, K. Steward, and V. Vuglinski. 2000. Historical trends in lake and river ice cover in the Northern Hemisphere. Science 289:1743–1746.

Magnuson, J.J., B.J. Benson, O.P. Jensen, T.B. Clark, V. Card, M.N. Futter, P.A. Soranno, and K.M. Stewart. 2005. Persistence of coherence of ice-off dates for inland lakes across the Laurentian Great Lakes region. Verh. Internat. Verein. Limnol. 29:521–527.

Ice Breakup in Two Alaskan Rivers

Identification

1. Description

This regional feature highlights the annual date of river ice breakup for two rivers: the Tanana River at Nenana, Alaska, and the Yukon River at Dawson City, Yukon Territory, Canada (the first town upstream from the Alaskan border). These data are available from 1917 (Tanana) and 1896 (Yukon) to present. The date of ice breakup is affected by several environmental factors, including air temperature, precipitation, wind, and water temperature. Tracking the date of ice breakup over time can provide important information about how the climate is changing at a more localized scale. Changes in this date can pose significant socioeconomic, geomorphic, and ecologic consequences (Beltaos and Burrell, 2003).

2. Revision History

December 2013:   Feature proposed.
May 2014:   Updated with data through 2014.

Data Sources

3. Data Sources

This feature presents the annual ice breakup data collected as part of the Nenana Ice Classic and Yukon River Breakup competitions. The Nenana Ice Classic is an annual competition to guess the exact timing of the breakup of ice in the Tanana River. Since its inception in 1917, the competition has paid more than $11 million in winnings, with a jackpot of $363,627 in 2014. A similar betting tradition occurs with the Yukon River in Dawson City, where ice breakup dates have been recorded since 1896.

4. Data Availability

All of the ice breakup data used are publicly available. Nenana Ice Classic data from 1917 to 2003 come from the National Snow and Ice Data Center (NSIDC), which maintains a comprehensive database at: http://nsidc.org/data/lake_river_ice. Data from 2004 to present have not yet been added to the NSIDC database, so these data were obtained directly from the Nenana Ice Classic organization; see the "Brochure" tab at: www.nenanaakiceclassic.com. Ice breakup dates from 1896 to present for the Yukon River are maintained by Mammoth Mapping and are available at: http://yukonriverbreakup.com/statistics.html.

Methodology

5. Data Collection

To measure the exact time of ice breakup, residents in Nenana and Dawson City use tripods placed on the ice in the center of the river. This tripod is attached by a cable to a clock on the shore, so that when the ice under the tripod breaks or starts to move, the tripod will move and pull the cable, stopping the clock with the exact date and time of the river ice breakup. In Nenana, the same wind-up clock has been used since the 1930s. Prior to the tripod method, observers watched from shore for movement of various objects placed on the ice. Dawson City also used onshore observers watching objects on the ice during the early years of the competition. For more information about these competitions, see: www.nenanaakiceclassic.com and: http://yukonriverbreakup.com/index.html.

6. Derivation

Figure 1 plots the annual ice breakup dates for each river. For some years, the original data set included the exact time of day when the ice broke, which could allow dates to be expressed as decimals (e.g., 120.5 would be noon on Julian day 120, which is the 120th day of the year). However, some other years in the data set did not include a specific time. Thus, for consistency, EPA chose to plot and analyze integer dates for all years (e.g., the example above would simply be treated as day #120).

Some data points were provided in the form of Julian days. In other cases where data points were provided in the form of calendar dates (e.g., May 1), EPA converted them to Julian days following the same method that was used to calculate Julian days in the original data set. By this method, January 1 = day 1, etc. The method also accounts for leap years, such that April 30 = day 120 in a non-leap year and day 121 in a leap year, for example. Figure 1 actually plots Julian dates, but the corresponding non-leap year calendar dates have been added to the y-axis to provide a more familiar frame of reference. This means that an ice breakup date of April 30 in a leap year will actually be plotted at the same level as May 1 from a non-leap year, for example, and it will appear to be plotted at May 1 with respect to the y-axis.

No annual data points were missing in the periods of record for these two rivers. This feature does not attempt to portray data beyond the time periods of observation.

7. Quality Assurance and Quality Control

The method of measuring river ice breakup ensures that an exact date and time is captured. Furthermore, the heavy betting tradition at both locations has long ensured a low tolerance for errors, as money is at stake for the winners and losers.

Analysis

8. Comparability Over Time and Space

River ice breakup dates have been recorded annually for the Tanana River since 1917 and for the Yukon River since 1896, using a measuring device or other objects placed on the river ice at the same location every year. This consistency allows for comparability over time.

9. Data Limitations

Factors that may impact the confidence, application, or conclusions drawn from the data are as follows:

1. While the record of river ice breakup dates is comprehensive, there are no corresponding environmental measurements (e.g., water conditions, air temperature), which limits one's ability to directly connect changes in river ice breakup to changes in climate.

2. Other factors, such as local development and land use patterns, may also affect the date of ice breakup. However, the two locations featured here are fairly remote and undeveloped, so the ice breakup dates are more likely to reflect natural changes in weather and climate conditions.

10. Sources of Uncertainty

This regional feature is likely to have very little uncertainty. The measurements are simple (i.e., the day when the ice starts to move at a particular location) and are collected with a device rather than relying on the human eye. Measurements have followed a consistent approach over time, and the competitive nature of the data collection effort means it is highly visible and transparent to the community, with low tolerance for error.

11. Sources of Variability

Natural climatic and hydrologic variations are likely to create year-to-year variation in ice breakup dates. For a general idea of the variability inherent in these types of time series, see Magnuson et al. (2000) and Jensen et al. (2007)—two papers that discuss variability and statistical significance for a broader set of lakes and rivers.

12. Statistical/Trend Analysis

EPA calculated long-term trends in river ice breakup for the Tanana and Yukon rivers by ordinary least-squares linear regression to support statements in the "Key Points" text. Both long-term trends were statistically significant at a 95 percent confidence level:

- Tanana regression slope, 1917–2014: -0.070 days/year (p = 0.001).
- Yukon regression slope, 1896–2014: -0.054 days/year (p < 0.001).

Both of these regressions are based on Julian dates, so they account for the influence of leap years (see Section 6 for more discussion of leap years). These regressions are also based on integer values for all years. As described in Section 6, some of the available data points included time of day, but others did not, so the graph and the regression analysis use integer dates for consistency.

References

Beltaos, S., and B.C. Burrell. 2003. Climatic change and river ice breakup. Can. J. Civil Eng. 30:145–155.

Jensen, O.P., B.J. Benson, and J.J. Magnuson. 2007. Spatial analysis of ice phenology trends across the Laurentian Great Lakes region during a recent warming period. Limnol. Oceanogr. 52(5):2013–2026.

Magnuson, J., D. Robertson, B. Benson, R. Wynne, D. Livingstone, T. Arai, R. Assel, R. Barry, V. Card, E. Kuusisto, N. Granin, T. Prowse, K. Steward, and V. Vuglinski. 2000. Historical trends in lake and river ice cover in the Northern Hemisphere. Science 289:1743–1746.

Snowfall

Identification

1. Indicator Description

Warmer temperatures associated with climate change can influence snowfall by altering weather patterns, causing more precipitation overall, and causing more precipitation to fall in the form of rain instead of snow. Thus, tracking metrics of snowfall over time can provide a useful perspective on how the climate may be changing aspects of precipitation. This indicator examines how snowfall has changed across the contiguous 48 states over time.

Components of this indicator include:

- Trends in total winter snowfall accumulation in the contiguous 48 states since 1930 (Figure 1).
- Changes in the ratio of snowfall to total winter precipitation since 1949 (Figure 2).

2. Revision History

December 2011:   Indicator developed.
May 2012:   Updated Figure 2 with data through 2011.
April 2014:   Updated Figure 2 with data through 2014.

Data Sources

3. Data Sources

The data used for this indicator are based on two studies published in the peer-reviewed literature: Kunkel et al. (2009) (Figure 1) and an update to Feng and Hu (2007) (Figure 2). Both studies are based on long-term weather station records compiled by the National Oceanic and Atmospheric Administration's (NOAA's) National Climatic Data Center (NCDC).

4. Data Availability

Figure 1. Change in Total Snowfall in the Contiguous 48 States, 1930–2007

EPA acquired Figure 1 data directly from Dr. Kenneth Kunkel of NOAA's Cooperative Institute for Climate and Satellites (CICS). Kunkel's analysis is based on data from weather stations that are part of NOAA's Cooperative Observer Program (COOP). Complete data, embedded definitions, and data descriptions for these stations can be found online at: www.ncdc.noaa.gov/doclib. State-specific data can be found at: www7.ncdc.noaa.gov/IPS/coop/coop.html;jsessionid=312EC0892FFC2FBB78F63D0E3ACF6CBC. There are no confidentiality issues that may limit accessibility. Additional metadata can be found at: www.nws.noaa.gov/om/coop.

Figure 2. Change in Snow-to-Precipitation Ratio in the Contiguous 48 States, 1949–2014

EPA acquired data from the U.S. Historical Climatology Network (USHCN), a compilation of weather station data maintained by NOAA. The USHCN allows users to download daily or monthly data at: www.ncdc.noaa.gov/oa/climate/research/ushcn. This website also provides data descriptions and other metadata. The data were taken from USHCN Version 2.

Methodology

5. Data Collection

Systematic collection of weather data in the United States began in the 1800s. Since then, observations have been recorded at 23,000 different stations. At any given time, observations are recorded at approximately 8,000 stations.

NOAA's National Weather Service (NWS) operates some stations (called first-order stations), but the vast majority of U.S. weather stations are part of the COOP network, which represents the core climate network of the United States (Kunkel et al., 2005). Cooperative observers include state universities, state and federal agencies, and private individuals. Observers are trained to collect data following NWS protocols, and equipment to gather these data is provided and maintained by the NWS.

Data collected by COOP are referred to as U.S. Daily Surface Data or Summary of the Day data. General information about the NWS COOP data set is available at: www.nws.noaa.gov/os/coop/what-is-coop.html. Sampling procedures are described in the full metadata for the COOP data set available at: www.nws.noaa.gov/om/coop. For more information about specific instruments and how they work, see: www.nws.noaa.gov/om/coop/training.htm.

NCDC also maintains the USHCN, which contains data from a subset of COOP and first-order weather stations that meet certain selection criteria and undergo additional levels of quality control. USHCN contains precipitation data from approximately 1,200 stations within the contiguous 48 states. The period of record varies for each station but generally includes most of the 20th century. One of the objectives in establishing the USHCN was to detect secular changes in regional rather than local climate. Therefore, stations included in this network are only those believed to not be influenced to any substantial degree by artificial changes of local environments. To be included in the USHCN, a station had to meet certain criteria for record longevity, data availability (percentage of available values), spatial coverage, and consistency of location (i.e., experiencing few station changes). An additional criterion, which sometimes compromised the preceding criteria, was the desire to have a uniform distribution of stations across the United States. Included with the data set are metadata files that contain information about station moves, instrumentation, observing times, and elevation. NOAA's website (www.ncdc.noaa.gov/oa/climate/research/ushcn) provides more information about USHCN data collection.

Figure 1. Change in Total Snowfall in the Contiguous 48 States, 1930–2007

The analysis in Figure 1 is based on snowfall (in inches), which weather stations measure daily through manual observation using a snow measuring rod. The measuring rod is a stick that observers use to measure the depth of snow.

The study on which this indicator is based includes data from 419 COOP stations in the contiguous United States for the months of October to May. These stations were selected using screening criteria that were designed to identify stations with the most consistent methods and most reliable data over time. Screening criteria are described in greater detail in Section 7.

Figure 2. Change in Snow-to-Precipitation Ratio in the Contiguous 48 States, 1949–2014

The analysis in Figure 2 is based on snowfall and precipitation measurements collected with standard gauges that "catch" precipitation, thus allowing weather stations to report daily precipitation totals. These gauges catch both solid (snow) and liquid (rain) precipitation. At each station, total daily precipitation is reported as a liquid equivalent depth based on one of two types of calculations: 1) precipitation is melted and the liquid depth measured, or 2) the precipitation is weighed. These methods are described by Huntington et al. (2004) and Knowles et al. (2006). Some stations occasionally use snow depth to calculate liquid equivalent by assuming a 10:1 density ratio between snow and rain. However, stations using this method extensively were excluded from the analysis because the assumed ratio does not always hold true, and because such an assumption could introduce a bias if snow density were also changing over time. Indeed, other analyses have cited changes in the density of snow over time, as warmer conditions lead to denser snow, particularly in late winter and early spring (Huntington et al., 2004).

This study uses data from 261 USHCN stations in the contiguous United States. Stations south of 37°N latitude were not included because most of them receive minimal amounts of snow each year. Additional site selection criteria are described in Section 7. This analysis covers the months from November through March, and each winter has been labeled based on the year in which it ends. For example, the data for "2014" represent the season that extended from November 2013 through March 2014.

6. Indicator Derivation

Figure 1. Change in Total Snowfall in the Contiguous 48 States, 1930–2007

At each station, daily snowfall totals have been summed to get the total snowfall for each winter. Thus, this figure technically reports trends from the winter of 1930–1931 to the winter of 2006–2007. Long-term trends in snowfall accumulation for each station are derived using an ordinary least-squares linear regression of the annual totals. Kunkel et al. (2009) describe analytical procedures in more detail. The lead author of Kunkel et al. (2009) conducted the most recent version of this analysis for EPA.

Figure 2. Change in Snow-to-Precipitation Ratio in the Contiguous 48 States, 1949–2014

EPA developed Figure 2 by following an approach published by Feng and Hu (2007). Using precipitation records from the USHCN Version 2, EPA calculated a snow-to-precipitation (S:P) ratio for each year by comparing the total snowfall during the months of interest (in terms of liquid-water equivalent) with total precipitation (snow plus rain). Long-term rates of change at each station were determined using a Kendall's tau slope estimator. This method of statistical analysis is described in Sen (1968) and Gilbert (1987). For a more detailed description of analytical methods, see Feng and Hu (2007).

7. Quality Assurance and Quality Control

The NWS has documented COOP methods, including training manuals and maintenance of equipment, at: www.nws.noaa.gov/os/coop/training.htm. These training materials also discuss quality control of the underlying data set. Additionally, pre-1948 data in the COOP data set have recently been digitized from hard copy. Quality control procedures associated with digitization and other potential sources of error are discussed in Kunkel et al. (2005).

Quality control procedures for USHCN Version 1 are summarized at: www.ncdc.noaa.gov/oa/climate/research/ushcn/ushcn.html#QUAL. Homogeneity testing and data correction methods are described in numerous peer-reviewed scientific papers by NCDC. Quality control procedures for USHCN Version 2 are summarized at: www.ncdc.noaa.gov/oa/climate/research/ushcn/#processing.

Figure 1. Change in Total Snowfall in the Contiguous 48 States, 1930–2007

Kunkel et al. (2009) filtered stations for data quality by selecting stations with records that were at least 90 percent complete over the study period. In addition, each station must possess at least five years of records during the decade at either end of the trend analysis (i.e., 1930s and 2000s) because data near the endpoints exert a relatively heavy influence on the overall trend. Year-to-year statistical outliers were also extensively cross-checked against nearby stations or *Climatological Data* publications when available. Any discrepancies with apparent regional trends were reviewed and evaluated by a panel of seven climate experts for data quality assurance. A more extensive description of this process, along with other screening criteria, can be found in Kunkel et al. (2009).

Figure 2. Change in Snow-to-Precipitation Ratio in the Contiguous 48 States, 1949–2014

Following the methods outlined by Feng and Hu (2007), EPA applied a similar filtering process to ensure data quality and consistency over time. Stations missing certain amounts of snow or precipitation data per month or per season were excluded from the study. Additional details about quality assurance are described in Feng and Hu (2007).

With assistance from the authors of Feng and Hu (2007), EPA added another screening criterion in 2012 that excluded sites that frequently used a particular estimation method to calculate snow water equivalent. This resulted in 85 fewer stations compared with the dataset in Feng and Hu (2007). Specifically, instructions given to observers in the early to mid-twentieth century provided an option to convert the measured snowfall to precipitation using a 10:1 ratio if it was impractical to melt the snow. Many observers have used this option in their reports of daily precipitation, although the number of observers using this option has declined through the years. The actual snowfall-to-liquid precipitation density ratio is related to factors such as air temperature during the snow event; the ratio varies spatially and it may be changing over time (e.g., Huntington et al., 2004). The median ratio in recent decades has been approximately 13:1 in the contiguous United States (Baxter et al., 2005; Kunkel et al., 2007), which suggests that using a 10:1 ratio could generally overestimate daily precipitation. Total winter precipitation in a snowy climate would thus be problematic if a large portion of the daily precipitation was estimated using this ratio, and apparent changes in S:P ratio over time could be biased if the average density of snow were also changing over time. To reduce the impact of this practice on the results, this analysis excluded records where winter (November to March) had more than 10 days with snowfall depth larger than 3.0 cm and where more than 50 percent of those snowy days reported total precipitation using the 10:1 ratio.

EPA also reviewed other analyses of snow-to-precipitation ratios. Knowles et al. (2006) used substantially different site selection criteria from those used for EPA's indicator. The study underlying EPA's indicator, Feng and Hu (2007), uses stricter criteria that exclude several stations in the Southern Rockies and other higher elevation sites. Further, Knowles et al. (2006) relied on an older version of the methods than what was used for EPA's indicator.

Analysis

8. Comparability Over Time and Space

Techniques for measuring snow accumulation and precipitation are comparable over space and time, as are the analytical methods that were used to develop Figures 1 and 2. Steps have been taken to remove stations where trends could be biased by changes in methods, location, or surrounding land cover.

9. Data Limitations

Factors that may impact the confidence, application, or conclusions drawn from this indicator are as follows:

1. While steps have been taken to limit this indicator to weather stations with the most consistent methods and the highest-quality data, several factors make it difficult to measure snowfall precisely. The snow accumulations shown in Figure 1 are based on the use of measuring rods. This measurement method is subject to human error, as well as the effects of wind (drifting snow) and the surrounding environment (such as tall trees). Similarly, precipitation gauges for Figure 2 may catch less snow than rain because of the effects of wind. This indicator has not been corrected for gauge catch efficiency. However, a sensitivity analysis described by Knowles at el. (2006) found that undercatch should have relatively little effect on overall trends in S:P ratios over time. It is not possible to account for gauge catch efficiency precisely because station-specific gauge efficiency assessments are generally unavailable (Knowles et al., 2006).

2. Both figures are limited to the winter season. Figure 1 comes from an analysis of October-to-May snowfall, while Figure 2 covers November through March. Although these months account for the vast majority of snowfall in most locations, this indicator might not represent the entire snow season in some areas.

3. Taken by itself, a decrease in S:P ratio does not necessarily mean that a location is receiving less snow than it used to or that snow has changed to rain. For example, a station with increased rainfall in November might show a decline in S:P ratio even with no change in snowfall during the rest of the winter season. This example illustrates the value of examining snowfall trends from multiple perspectives, as this indicator seeks to do.

4. Selecting only those stations with high-quality long-term data leads to an uneven density of stations for this indicator. Low station density limits the conclusions that can be drawn about certain regions such as the Northeast and the Intermountain West.

5. Most of the data shown for mountainous regions come from lower elevations (towns in valleys) because that is where permanent COOP weather stations tend to be located. Thus, the results are not necessarily representative of higher elevations, which might not have the same sensitivity to temperature change as lower elevations. Another monitoring network, called SNOTEL, measures snow depth at higher-elevation sites. SNOTEL data are an important part of EPA's Snowpack indicator. SNOTEL sites are limited to mountainous areas of the West, however—none in the East—and they do not measure daily rainfall, which is necessary for the analysis in Figure 2. Thus, EPA has not included SNOTEL data in this indicator.

10. Sources of Uncertainty

Quantitative estimates of uncertainty are not available for Figure 1, Figure 2, or most of the underlying measurements.

Figure 1. Change in Total Snowfall in the Contiguous 48 States, 1930–2007

Snow accumulation measurements are subject to human error. Despite the vetting of observation stations, some error could also result from the effects of wind and surrounding cover, such as tall trees. Some records have evidence of reporting errors related to missing data (i.e., days with no snow being reported as missing data instead of "0 inches"), but Kunkel et al. (2009) took steps to correct this error in cases where other evidence (such as daily temperatures) made it clear that an error was present.

Figure 2. Change in Snow-to-Precipitation Ratio in the Contiguous 48 States, 1949–2014

The source study classifies all precipitation as "snow" for any day that received some amount of snow. This approach has the effect of overestimating the amount of snow during mixed snow-sleet-rain conditions. Conversely, wind effects that might prevent snow from settling in gauges will tend to bias the S:P ratio toward rainier conditions. However, Section 9 explains that gauge catch efficiency should not substantially affect the conclusions that can be drawn from this indicator.

11. Sources of Variability

Snowfall naturally varies from year to year as a result of typical variation in weather patterns, multi-year climate cycles such as the El Niño–Southern Oscillation and Pacific Decadal Oscillation (PDO), and other factors. The PDO switches between "warm" and "cool" phases approximately every 20 to 30 years (see: http://jisao.washington.edu/pdo and publications cited therein), so the 50+-year record shown in this indicator may be affected by a few PDO phase transitions. Overall, though, the length of data available for this indicator should support a reliable analysis of long-term trends.

Snowfall is influenced by temperature and a host of other factors such as regional weather patterns, local elevation and topography, and proximity to large water bodies. These differences can lead to great variability in trends among stations—even stations that may be geographically close to one another.

12. Statistical/Trend Analysis

Figure 1. Change in Total Snowfall in the Contiguous 48 States, 1930–2007

This indicator reports a trend for each station based on ordinary least-squares linear regression. The significance of each station's trend was not reported in Kunkel et al. (2009).

Figure 2. Change in Snow-to-Precipitation Ratio in the Contiguous 48 States, 1949–2014

Feng and Hu (2007) calculated a long-term trend in S:P ratio at each station using the Kendall's tau method. EPA used the same method for the most recent data update. EPA also determined a z-score for every station. Based on these z-scores, Figure 2 identifies which station trends are statistically significant based on a 95 percent confidence threshold (i.e., a z-score with an absolute value greater than 1.645).

References

Baxter, M.A., C.E. Graves, and J.T. Moore. 2005. A climatology of snow-to-liquid ratio for the contiguous United States. Weather Forecast. 20:729–744.

Feng, S., and Q. Hu. 2007. Changes in winter snowfall/precipitation ratio in the contiguous United States. J. Geophys. Res. 112:D15109.

Gilbert, R.O. 1987. Statistical methods for environmental pollution monitoring. New York, NY: Van Nostrand Reinhold.

Huntington, T.G., G.A. Hodgkins, B.D. Keim, and R.W. Dudley. 2004. Changes in the proportion of precipitation occurring as snow in New England (1949–2000). J. Climate 17:2626–2636.

Knowles, N., M.D. Dettinger, and D.R. Cayan. 2006. Trends in snowfall versus rainfall in the western United States. J. Climate 19:4545–4559.

Kunkel, K.E., D.R. Easterling, K. Hubbard, K. Redmond, K. Andsager, M.C. Kruk, and M.L. Spinar. 2005. Quality control of pre-1948 Cooperative Observer Network data. J. Atmos. Ocean. Tech. 22:1691–1705.

Kunkel, K.E., M. Palecki, K.G. Hubbard, D.A. Robinson, K.T. Redmond, and D.R. Easterling. 2007. Trend identification in twentieth-century U.S. snowfall: The challenges. J. Atmos. Ocean. Tech. 24:64–73.

Kunkel, K.E., M. Palecki, L. Ensor, K.G. Hubbard, D. Robinson, K. Redmond, and D. Easterling. 2009. Trends in twentieth-century U.S. snowfall using a quality-controlled dataset. J. Atmos. Ocean. Tech. 26:33–44.

Sen, P.K. 1968. Estimates of the regression coefficient based on Kendall's tau. J. Am. Stat. Assoc. 63:1379–1389.

Snow Cover

Identification

1. Indicator Description

This indicator measures changes in the amount of land in North America covered by snow. The amount of land covered by snow at any given time is influenced by climate factors such as the amount of snowfall an area receives, the timing of that snowfall, and the rate of melting on the ground. Thus, tracking snow cover over time can provide a useful perspective on how the climate may be changing. Snow cover is also climatically meaningful because it exerts an influence on climate through the albedo effect (i.e., the color and reflectivity of the Earth's surface).

Components of this indicator include:

- Average annual snow cover since 1972 (Figure 1).
- Average snow cover by season since 1972 (Figure 2).

2. Revision History

April 2010:	Indicator posted.
January 2012:	Updated with data through 2011.
February 2012:	Expanded to include snow cover by season (new Figure 2).
August 2013:	Updated on EPA's website with data through 2012.
March 2014:	Updated with data through 2013.

Data Sources

3. Data Sources

This indicator is based on a Rutgers University Global Snow Lab (GSL) reanalysis of digitized maps produced by the National Oceanic and Atmospheric Administration (NOAA) using their Interactive Multisensor Snow and Ice Mapping System (IMS).

4. Data Availability

Complete weekly and monthly snow cover extent data for North America (excluding Greenland) are publicly available for users to download from the GSL website at: http://climate.rutgers.edu/snowcover/table_area.php?ui_set=2. A complete description of these data can be found on the GSL website at: http://climate.rutgers.edu/snowcover/index.php.

The underlying NOAA gridded maps are also publicly available. To obtain these maps, visit the NOAA IMS website at: www.natice.noaa.gov/ims.

Methodology

5. Data Collection

This indicator is based on data from instruments on polar-orbiting satellites, which orbit the Earth continuously and are able to map the entire surface of the Earth. These instruments collect images that can be used to generate weekly maps of snow cover. Data are collected for the entire Northern Hemisphere; this indicator includes data for all of North America, excluding Greenland.

Data were compiled as part of NOAA's IMS, which incorporates imagery from a variety of satellite instruments (Advanced Very High Resolution Radiometer [AVHRR], Geostationary Satellite Server [GOES], Special Sensor Microwave Imager [SSMI], etc.) as well as derived mapped products and surface observations. Characteristic textured surface features and brightness allow for snow to be identified and data to be collected on the percentage of snow cover and surface albedo (reflectivity) (Robinson et al., 1993).

NOAA's IMS website (www.natice.noaa.gov/ims) lists peer-reviewed studies and websites that discuss the data collection methods, including the specific satellites that have provided data at various times. For example, NOAA sampling procedures are described in Ramsay (1998). For more information about NOAA's satellites, visit: www.nesdis.noaa.gov/about_satellites.html.

6. Indicator Derivation

NOAA digitizes satellite maps weekly using the National Meteorological Center Limited-Area Fine Mesh grid. In the digitization process, an 89-by-89-cell grid is placed over the Northern Hemisphere and each cell has a resolution range of 16,000 to 42,000 square kilometers. NOAA then analyzes snow cover within each of these grid cells.

Rutgers University's GSL reanalyzes the digitized maps produced by NOAA to correct for biases in the data set caused by locations of land masses and bodies of water that NOAA's land mask does not completely resolve. Initial reanalysis produces a new set of gridded data points based on the original NOAA data points. Both original NOAA data and reanalyzed data are filtered using a more detailed land mask produced by GSL. These filtered data are then used to make weekly estimates of snow cover. GSL determines the weekly extent of snow cover by placing an 89-by-89-cell grid over the Northern Hemisphere snow cover map and calculating the total area of all grid cells that are at least 50 percent snow-covered. GSL generates monthly maps based on an average of all weeks within a given month. Weeks that straddle the end of one month and the start of another are weighted proportionally.

EPA obtained weekly estimates of snow-covered area and averaged them to determine the annual average extent of snow cover in square kilometers. EPA obtained monthly estimates of snow-covered area to determine the seasonal extent of snow cover in square kilometers. For each year, a season's extent was determined by averaging the following months:

- Winter: December (of the prior calendar year), January, and February.
- Spring: March, April, and May.
- Summer: June, July, and August.
- Fall: September, October, and November.

EPA converted all of these values to square miles to make the results accessible to a wider audience.

NOAA's IMS website describes the initial creation and digitization of gridded maps; see: www.natice.noaa.gov/ims. The GSL website provides a complete description of how GSL reanalyzed NOAA's gridded maps to determine weekly and monthly snow cover extent. See: http://climate.rutgers.edu/snowcover/docs.php?target=vis and http://climate.rutgers.edu/snowcover/docs.php?target=cdr. Robinson et al. (1993) describe GSL's methods, while Helfrich et al. (2007) document how GSL has accounted for methodological improvements over time. All maps were recently reanalyzed using the most precise methods available, making this the best available data set for assessing snow cover on a continental scale.

7. Quality Assurance and Quality Control

Quality assurance and quality control (QA/QC) measures occur throughout the analytical process, most notably in the reanalysis of NOAA data by GSL. GSL's filtering and correction steps are described online (http://climate.rutgers.edu/snowcover/docs.php?target=vis) and in Robinson et al. (1993). Ramsey (1998) describes the validation plan for NOAA digitized maps and explains how GSL helps to provide objective third-party verification of NOAA data.

Analysis

8. Comparability Over Time and Space

Steps have been taken to exclude less reliable early data from this indicator. Although NOAA satellites began collecting snow cover imagery in 1966, early maps had a lower resolution than later maps (4 kilometers versus 1 kilometer in later maps) and the early years also had many weeks with missing data. Data collection became more consistent with better resolution in 1972, when a new instrument called the Very High Resolution Radiometer (VHRR) came online. Thus, this indicator presents only data from 1972 and later.

Mapping methods have continued to evolve since 1972. Accordingly, GSL has taken steps to reanalyze older maps to ensure consistency with the latest approach. GSL provides more information about these correction steps at: http://climate.rutgers.edu/snowcover/docs.php?target=cdr.

Data have been collected and analyzed using consistent methods over space. The satellites that collect the data cover all of North America in their orbital paths.

9. Data Limitations

Factors that may impact the confidence, application, or conclusions drawn from this indicator are as follows:

1. Satellite data collection is limited by anything that obscures the ground, such as low light conditions at night, dense cloud cover, or thick forest canopy. Satellite data are also limited by difficulties discerning snow cover from other similar-looking features such as cloud cover.

2. Although satellite-based snow cover totals are available starting in 1966, some of the early years are missing data from several weeks (mainly during the summer), which would lead to an inaccurate annual or seasonal average. Thus, the indicator is restricted to 1972 and later, with all years having a full set of data.

3. Discontinuous (patchy) snow cover poses a challenge for measurement throughout the year, particularly with spectrally and spatially coarse instruments such as AVHRR (Molotch and Margulis, 2008).

4. Summer snow mapping is particularly complicated because many of the patches of snow that remain (e.g., high in a mountain range) are smaller than the pixel size for the analysis. This leads to reduced confidence in summer estimates. When summer values are incorporated into an annual average, however, variation in summer values has relatively minimal influence on the overall results.

10. Sources of Uncertainty

Uncertainty measurements are not readily available for this indicator or for the underlying data. Although exact uncertainty estimates are not available, extensive QA/QC and third-party verification measures show that steps have been taken to minimize uncertainty and ensure that users are able to draw accurate conclusions from the data. Documentation available from GSL (http://climate.rutgers.edu/snowcover/docs.php?target=vis) explains that since 1972, satellite mapping technology has had sufficient accuracy to support continental-scale climate studies. Although satellite data have some limitations (see Section 9), maps based on satellite imagery are often still superior to maps based on ground observations, which can be biased due to sparse station coverage—both geographically and in terms of elevation (e.g., a station in a valley will not necessarily have snow cover when nearby mountains do)—and by the effects of urban heat islands in locations such as airports. Hence, satellite-based maps are generally more representative of regional snow extent, particularly for mountainous or sparsely populated regions.

11. Sources of Variability

Figures 1 and 2 show substantial year-to-year variability in snow cover. This variability naturally results from variation in weather patterns, multi-year climate cycles such as the El Niño–Southern Oscillation and Pacific Decadal Oscillation, and other factors. Underlying weekly measurements have even more variability. This indicator accounts for these factors by presenting a long-term record (several decades) and calculating annual and seasonal averages.

Generally, decreases in snow cover duration have been most pronounced along mid-latitude continental margins where seasonal mean air temperatures range from -5 to +5°C (Brown and Mote, 2009).

12. Statistical/Trend Analysis

EPA performed an initial assessment of trends in square miles (mi^2) per year using ordinary least-squares linear regression, which led to the following results:

- Annual average, 1972–2013: -3,527 mi^2/year (p = 0.038).
- Winter, 1972–2013: +3,715 mi^2/year (p = 0.23).

- Spring, 1972–2013: -6,649 mi^2/year (p = 0.034).
- Summer, 1972–2013: -14,601 mi^2/year (p < 0.001).
- Fall, 1972–2013: +2,683 mi^2/year (p = 0.30).

Thus, long-term linear trends in annual average, spring, and summer snow cover are all significant to a 95 percent level (p < 0.05), while winter and fall trends are not. To conduct a more complete analysis would potentially require consideration of serial correlation and other more complex statistical factors.

References

Brown, R.D., and P.W. Mote. 2009. The response of Northern Hemisphere snow cover to a changing climate. J. Climate 22:2124–2145.

Helfrich, S.R., D. McNamara, B.H. Ramsay, T. Baldwin, and T. Kasheta. 2007. Enhancements to, and forthcoming developments in the Interactive Multisensor Snow and Ice Mapping System (IMS). Hydrol. Process. 21:1576–1586.

Molotch, N.P., and S.A. Margulis. 2008. Estimating the distribution of snow water equivalent using remotely sensed snow cover data and a spatially distributed snowmelt model: A multi-resolution, multi-sensor comparison. Adv. Water Resour. 31(11):1503–1514.

Ramsay, B.H. 1998. The Interactive Multisensor Snow and Ice Mapping System. Hydrol. Process. 12:1537–1546. www.natice.noaa.gov/ims/files/ims-hydro-proc-ramsay-1998.pdf.

Robinson, D.A., K.F. Dewey, and R.R. Heim, Jr. 1993. Global snow cover monitoring: An update. B. Am. Meteorol. Soc. 74:1689–1696.

Snowpack

Identification

1. Indicator Description

This indicator describes changes in springtime mountain snowpack in the western United States between 1955 and 2013. Mountain snowpack is a key component of the water cycle in the western United States, storing water in the winter when the snow falls and releasing it in spring and early summer when the snow melts. Changes in snowpack over time can reflect a changing climate, as temperature and precipitation are key factors that influence the extent and depth of snowpack. In a warming climate, more precipitation will be expected to fall as rain rather than snow in most areas—reducing the extent and depth of snowpack. Higher temperatures in the spring can cause snow to melt earlier.

2. Revision History

April 2010:   Indicator posted.
April 2014:   Updated with data through 2013.

Data Sources

3. Data Sources

This indicator is based largely on data collected by the U.S. Department of Agriculture's (USDA's) Natural Resources Conservation Service (NRCS). Additional snowpack data come from observations made by the California Department of Water Resources.

4. Data Availability

EPA obtained the data for this indicator from Dr. Philip Mote at Oregon State University. Dr. Mote had published an earlier version of this analysis (Mote et al., 2005) with data from about 1930 to 2000 and a map of trends from 1950 through 1997, and he and colleague Darrin Sharp were able to provide EPA with an updated analysis of trends from 1955 through 2013.

This analysis is based on snowpack measurements from NRCS and the California Department of Water Resources. Both sets of data are available to the public with no confidentiality or accessibility restrictions. NRCS data are available at: www.wcc.nrcs.usda.gov/snow/snotel-wedata.html. California data are available at: http://cdec.water.ca.gov/snow/current/snow/index.html. These websites also provide descriptions of the data. At the time of this analysis in April 2014, NRCS data were available through 2012; California data were available through 2013.

Methodology

5. Data Collection

This indicator uses snow water equivalent (SWE) measurements to assess trends in snowpack from 1955 through 2013. SWE is the amount of water contained within the snowpack at a particular location. It can be thought of as the depth of water that would result if the entire snowpack were to melt. Because snow can vary in density (depending on the degree of compaction, for example), converting to the equivalent amount of liquid water provides a more consistent metric than snow depth. Snowpack measurements have been extensively documented and have been used for many years to help forecast spring and summer water supplies, particularly in the western United States.

Snowpack data have been collected over the years using a combination of manual and automated techniques. All of these techniques are ground-based observations, as SWE is difficult to measure from aircraft or satellites—although development and validation of remote sensing for snowpack is a subject of ongoing research. Consistent manual measurements from "snow courses" or observation sites are available beginning in the 1930s, although a few sites started earlier. These measurements, typically taken near the first of each month between January and May or June, require an observer to travel to remote locations, on skis, snowshoes, snowmobile, or by helicopter, to measure SWE. At a handful of sites, an aircraft-based observer photographs snow depth against a permanent marker.

In 1979, NRCS and its partners began installing automated snowpack telemetry (SNOTEL) stations. Instruments at these stations automatically measure snowpack and related climatic data. The NRCS SNOTEL network now operates more than 650 remote sites in the western United States, including Alaska. In contrast to monthly manual snow course measurements, SNOTEL sensor data are recorded every 15 minutes and reported daily to two master stations. In most cases, a SNOTEL site was located near a snow course, and after a period of overlap to establish statistical relationships, the co-located manual snow course measurements were discontinued. However, hundreds of other manual snow course sites are still in use, and data from these sites are used to augment data from the SNOTEL network and provide more complete coverage of conditions throughout the western United States.

Additional snowpack data come from observations made by the California Department of Water Resources.

For information about each of the data sources and its corresponding sample design, visit the following websites:

- NRCS: www.wcc.nrcs.usda.gov/snow and www.wcc.nrcs.usda.gov/snowcourse.
- California Department of Water Resources:
  http://cdec.water.ca.gov/snow/info/DataCollecting.html.

The NRCS website describes both manual and telemetric snowpack measurement techniques in more detail at: www.wcc.nrcs.usda.gov/factpub/sect_4b.html. A training and reference guide for snow surveyors who use sampling equipment to measure snow accumulation is also available on the NRCS website at: www.wcc.nrcs.usda.gov/factpub/ah169/ah169.htm.

For consistency, this indicator examines trends at the same date each year. This indicator uses April 1 as the annual date for analysis because it is the most frequent observation date and it is extensively used for spring streamflow forecasting (Mote et al., 2005). Data are nominally attributed to April 1, but in reality, for some manually operated sites the closest measurement in a given year might have been collected slightly before or after April 1. The collection date is noted in the data set, and the California Department of Water Resources also estimates the additional SWE that would have accumulated between the collection date and April 1. For evaluating long-term trends, there is little difference between the two data sets.

This indicator focuses on the western United States (excluding Alaska) because this broad region has the greatest density of stations with long-term records. A total of 2,914 locations have recorded SWE measurements within the area of interest. This indicator is based on 701 stations with sufficient April 1 records spanning the period from 1955 through 2013.

The selection criteria for station inclusion in this analysis were as follows:

- The station must have data back to at least 1955.
- For the 10 years 2004–2013, the station must have at least five data points.
- Over the period 1955–2013, the station must have more April 1 SWE values greater than 0 than equal to 0.

In addition, stations were excluded even if they met all of the above requirements but still had large data gaps. Using these criteria, the minimum number of data points for any station in this analysis was 34. All but 13 stations had 50 or more data points.

6. Indicator Derivation

Dr. Mote's team calculated linear trends in April 1 SWE measurements from 1955 through 2013. For this indicator, 1955 was selected as a starting point because it is early enough to provide long records but late enough to include many sites in the Southwest where measurement began during the early 1950s. Trends were calculated for 1955 through 2013 at each snow course or SNOTEL location, and then these trends were converted to percent change since 1955. Note that this method can lead to an apparent loss exceeding 100 percent at a few sites (i.e., more than a 100 percent decrease in snowpack) in cases where the line of best fit passes through zero sometime before 2013, indicating that it is now most likely for that location to have no snowpack on the ground at all on April 1. It can also lead to large percentage increases for sites with a small initial value for the linear fit. For more details about the analytical procedures used to calculate trends and percent change for each location, see Mote et al. (2005).

EPA obtained a data file with coordinates and percent change for each station, and plotted the results on a map using ArcGIS software. Figure 1 shows trends at individual sites with measured data, with no attempt to generalize data over space.

7. Quality Assurance and Quality Control

Automated SNOTEL data are screened by computer to ensure that they meet minimum requirements before being added to the database. In addition, each automated data collection site receives maintenance and sensor adjustment annually. Data reliability is verified by ground truth measurements taken during regularly scheduled manual surveys, in which manual readings are compared with automated data to check that values are consistent. Based on these quality assurance and quality control (QA/QC) procedures, maintenance visits are conducted to correct deficiencies. Additional description of QA/QC procedures for the SNOTEL network can be found on the NRCS website at: www.wcc.nrcs.usda.gov/factpub/sect_4b.html.

QA/QC procedures for manual measurements by NRCS and by the California Department of Water Resources are largely unavailable online.

Additional QA/QC activities were conducted on the data obtained from NRCS and the California Department of Water Resources. Station data were checked for physically unrealistic values such as SWE larger than snow depth, or SWE or snow depth values far beyond the upper bounds of what would even be considered exceptional (i.e., 300 inches of snow depth or 150 inches of SWE). In these cases, after manual verification, suspect data were replaced with a "no data" value. In addition, the April-to-March ratio of SWE was evaluated, and any station that had a ratio greater than 10 was evaluated manually for data accuracy.

Analysis

8. Comparability Over Time and Space

For consistency, this indicator examines trends at the same point in time each year. This indicator uses April 1 as the annual date for analysis because it is the most frequent observation date and it is extensively used for spring streamflow forecasting (Mote et al., 2005). Data are nominally attributed to April 1, but, in reality, for some manually operated sites the closest measurement in a given year might have been collected as much as two weeks before or after April 1. However, in the vast majority of cases, the April 1 measurement was made within a few days of April 1.

Data collection methods have changed over time in some locations, particularly as automated devices have replaced manual measurements. However, agencies such as NRCS have taken careful steps to calibrate the automated devices and ensure consistency between manual and automatic measurements (see Section 7). They also follow standard protocols to ensure that methods are applied consistently over time and space.

9. Data Limitations

Factors that may impact the confidence, application, or conclusions drawn from this indicator are as follows:

1. EPA selected 1955 as a starting point for this analysis because many snow courses in the Southwest were established in the early 1950s, thus providing more complete spatial coverage. Some researchers have examined snowpack data within smaller regions over longer or shorter time frames and found that the choice of start date can make a difference in the magnitude of the resulting trends. For example, Mote et al. (2008) pointed out that lower-elevation snow courses in the Washington Cascades were mostly established after 1945, so limiting the analysis to sites established by 1945 results in a sampling bias toward higher, colder sites. They also found that starting the linear fit between 1945 and 1955—an unusually snowy period in the Northwest—led to somewhat larger average declines. Across the entire western United States,

though, the median percentage change and the percentage of sites with declines are fairly consistent, regardless of the start date.

2. Although most parts of the West have seen reductions in snowpack, consistent with overall warming trends, observed snowfall trends could be partially influenced by non-climatic factors such as observation methods, land use changes, and forest canopy changes. A few snow course sites have been moved over time—for example, because of the growth of recreational uses such as snowmobiling or skiing. Mote et al. (2005) also report that the mean date of "April 1" observations has grown slightly later over time.

10. Sources of Uncertainty

Uncertainty estimates are not readily available for this indicator or for the underlying snowpack measurements. However, the regionally consistent and in many cases sizable changes shown in Figure 1, along with independent hydrologic modeling studies (Mote et al., 2005; Ashfaq et al., 2013), strongly suggest that this indicator shows real secular trends, not simply the artifacts of some type of measurement error.

11. Sources of Variability

Snowpack trends may be influenced by natural year-to-year variations in snowfall, temperature, and other climate variables. To reduce the influence of year-to-year variability, this indicator looks at longer-term trends over the full 58-year time series.

Over a longer time frame, snowpack variability can result from variations in the Earth's climate or from non-climatic factors such as changes in observation methods, land use, and forest canopy.

12. Statistical/Trend Analysis

Figure 1 shows the results of a least-squares linear regression of annual observations at each individual site from 1955 through 2013. The statistical significance of each of these trends was examined using the Mann-Kendall test for significance and the Durbin-Watson test for serial correlation (autocorrelation). Of the 701 stations in this analysis, 159 had trends that were significant to a 95 percent level ($p < 0.05$) according to the Mann-Kendall test, with 13 of those sites showing autocorrelation (Durbin-Watson < 0.1). A block bootstrap (using both three- and five-year blocks) was applied to those 13 sites that had both significant autocorrelation and significant trends. In all cases, the Mann-Kendall test indicated significant trends ($p < 0.05$) even after applying the block bootstrap. As a result, it was determined that all 159 sites showed statistically significant trends: four with increases and 155 with decreases.

References

Ashfaq, M., S. Ghosh, S.-C. Kao, L.C. Bowling, P. Mote, D. Touma, S.A. Rauscher, and N.S. Diffenbaugh. 2013. Near-term acceleration of hydroclimatic change in the western U.S. J. Geophys. Res.-Atmos. 118(19): 10676–10693.

Mote, P.W., A.F. Hamlet, M.P. Clark, and D.P. Lettenmaier. 2005. Declining mountain snowpack in western North America. B. Am. Meteorol. Soc. 86(1):39–49.

Mote, P.W., A.F. Hamlet, and E.P. Salathé, Jr. 2008. Has spring snowpack declined in the Washington Cascades? Hydrol. Earth Syst. Sc. 12:193–206.

Heating and Cooling Degree Days

Identification

1. Indicator Description

This indicator measures trends in heating degree days (HDD) and cooling degree days (CDD) in the United States between 1895 and 2013. Heating and cooling degree days are measures that reflect the amount of energy needed to heat or cool a building to a comfortable temperature, given how cold or hot it is outside. A "degree day" indicates that the daily average outdoor temperature was one degree higher or lower than some comfortable baseline temperature (in this case, 65°F—a typical baseline used by the National Oceanic and Atmospheric Administration [NOAA]) on a particular day. The sum of the number of heating or cooling degree days over a year is roughly proportional to the annual amount of energy that would be needed to heat or cool a building in that location (Diaz and Quayle, 1980). Thus, HDD and CDD are rough surrogates for how climate change is likely to affect energy use for heating and cooling.

Components of this indicator include:

- Annual average HDD and CDD nationwide, compared with long-term averages (Figure 1).
- Change in annual HDD by state (Figure 2).
- Change in annual CDD by state (Figure 3).

2. Revision History

December 2013:   Indicator proposed.
January 2014:   Updated with data through 2013.

Data Sources

3. Data Sources

Data for this indicator were provided by NOAA's National Climatic Data Center (NCDC). These data are based on temperature measurements from weather stations overseen by NOAA's National Weather Service (NWS). These underlying data are maintained by NCDC.

4. Data Availability

EPA used data for this indicator that were posted to NCDC's website as of January 2014—specifically the product called "Time Bias Corrected Divisional Temperature-Precipitation-Drought Index," Series TD-9640. See: www1.ncdc.noaa.gov/pub/data/cirs/div-dataset-transition-readme.txt for a description how to access this data product and the successor products to which NCDC is currently transitioning.

Documentation for the time-bias-corrected temperature data on which the HDD and CDD series are based is available from NOAA (2002a). A description of the methodology used to generate HDD and CDD from monthly temperatures and standard deviations is available in NOAA (2002b) and NOAA (2002c).

NCDC maintains a set of databases that provide public access to daily and monthly temperature records from thousands of weather stations across the country. For access to these data and accompanying metadata, visit NCDC's website at: www.ncdc.noaa.gov. The weather stations used for this indicator are part of NOAA's Cooperative Observer Program (COOP). Complete data, embedded definitions, and data descriptions for these stations can be found online at: www.ncdc.noaa.gov/doclib. State-specific data can be found at: www7.ncdc.noaa.gov/IPS/coop/coop.html;jsessionid=312EC0892FFC2FBB78F63D0E3ACF6CBC. There are no confidentiality issues that may limit accessibility. Additional metadata can be found at: www.nws.noaa.gov/om/coop.

Methodology

5. Data Collection

This indicator measures annual average heating and cooling degree days nationwide and for each individual state. The HDD and CDD data are based on time-bias adjusted temperature data from COOP weather stations throughout the contiguous 48 states.

Systematic collection of weather data in the United States began in the 1800s. Since then, observations have been recorded from 23,000 stations. At any given time, observations are recorded from approximately 8,000 stations. COOP stations generally measure temperature at least hourly, and they record the maximum and minimum temperature for each 24-hour time span. Cooperative observers include state universities, state and federal agencies, and private individuals whose stations are managed and maintained by the NWS. Observers are trained to collect data following NWS protocols, and the NWS provides and maintains standard equipment to gather these data.

The COOP data set represents the core climate network of the United States (Kunkel et al., 2005). Data collected by COOP are referred to as U.S. Daily Surface Data or Summary of the Day data. Variables that are relevant to this indicator include observations of daily maximum and minimum temperatures. General information about the NWS COOP data set is available at: www.nws.noaa.gov/os/coop/what-is-coop.html. Sampling procedures are described in Kunkel et al. (2005) and in the full metadata for the COOP data set available at: www.nws.noaa.gov/om/coop.

6. Indicator Derivation

NCDC used several steps to calculate annual HDD and CDD data for each month of each year in each state (NOAA, 2002a,b,c).

First, the raw COOP station temperature data were adjusted to remove bias due to variation in the time of day at which temperature measurements were reported (Karl et al., 1986; NOAA, 2002a). This bias arises from the fact that, historically, some COOP stations have reported temperatures over climatological days ending at different times of day (e.g., over the 24-hour period ending at midnight versus the 24-hour period ending at 7:00 pm). This variation leads to different reported daily minimum and maximum temperatures, as well as inconsistencies in mean temperature (which historically has often been calculated as [*minimum temperature + maximum temperature*]/2). To address this problem, NCDC used the statistical adjustment procedure from Karl et al. (1986) to remove bias due to differences in time-of-day definitions. These biases were as large as 2°F in climate divisions within some states (NOAA, 2002a).

Second, the daily time-bias adjusted data were used to calculate mean temperatures in each month and year (NOAA, 2002b,c). Additionally, the data were used to calculate the standard deviation of daily temperatures in each location for each month (pooling across all years) over the entire period for which temperature data were available.

Third, NCDC estimated the total monthly heating and cooling degree days at each location. A crude way to find monthly totals would be to simply add all the daily HDD and CDD values over the course of the month. However, for reasons related to data quality, NCDC used a modified version of the procedure presented in Thom (1954a,b, 1966), which assumes that daily temperatures within a month are distributed normally. The expected number of HDD or CDD per month can then be expressed as a simple function of the actual monthly mean daily temperature and the long-term standard deviation of daily temperatures. The logic behind this approach is that HDD and CDD are measures that reflect both the mean (the "absolute value") and standard deviation (the "spread") of daily temperatures—and thus can be estimated from them. Although predictions based on this formula may be inaccurate for any particular day or week, on average across large time periods the predictions will be reasonably good. The rationale for using this approach is that daily COOP station data contain many "inhomogeneities" and missing data points that may add noise or bias to HDD and CDD estimates calculated directly from daily data. By estimating HDD and CDD following the Thom procedure, NCDC was able to generate estimates in a consistent way for all years of the data.

State and national averages for each year were calculated as follows:

1. NCDC calculated a monthly average HDD and CDD for each climate division (each state has up to 10 climate divisions; see: www.ncdc.noaa.gov/monitoring-references/maps/us-climate-divisions.php) by averaging the results from all stations within each division. All stations within a particular division were weighted equally.

2. NCDC calculated monthly averages for each state by weighting the climate divisions by their population. With this approach, state HDD and CDD values more closely reflect the conditions that the average resident of the state would experience.

3. NCDC calculated monthly averages for the contiguous 48 states by weighting the divisions or states according to their population.

4. NCDC and EPA added each year's monthly averages together to arrive at annual totals for the contiguous 48 states and for each individual state.

5. All population-based weighting was performed using population data from the 1990 U.S. Census. Figure 1 shows the national HDD and CDD averages as described above. EPA developed the maps of state-level changes in HDD and CDD (Figures 2 and 3) by separating the historical record into two periods of roughly equal length (1895–1953 and 1954–2013), then calculating how average annual HDD and CDD in each state changed between the two periods.

7. Quality Assurance and Quality Control

NOAA follows extensive quality assurance and quality control (QA/QC) procedures for collecting and compiling COOP weather station data. For documentation of COOP methods, including training manuals and maintenance of equipment, see: www.nws.noaa.gov/os/coop/training.htm. These training materials also discuss QC of the underlying data set. Additionally, pre-1948 data in the COOP data set have recently been digitized from hard copy. Quality control procedures associated with digitization and other potential sources of error are discussed in Kunkel et al. (2005).

As described in the previous section, NCDC took steps to ensure that time-of-day reporting biases were removed from the data, and that HDD and CDD were calculated using a methodology that generates results that can be compared over time.

Analysis

8. Comparability Over Time and Space

HDD and CDD have been calculated using the same methods for all locations and throughout the period of record. Each climate division contributes to the state and national averages in proportion to its population. All population-based weighting was performed using population data from the 1990 U.S. Census, so as to avoid ending up with an HDD or CDD trend line that reflects the influence of shifting populations (e.g., more people moving to areas with warmer climates).

9. Data Limitations

Factors that may impact the confidence, application, or conclusions drawn from this indicator are as follows:

1. Biases may have occurred as a result of changes over time in instrumentation, measuring procedures, and the exposure and location of the instruments. Where possible, data have been adjusted to account for changes in these variables. For more information on these corrections, see Section 7.

2. Observer errors, such as errors in reading instruments or writing observations on the form, are present in the earlier part of this data set. Additionally, uncertainty may be introduced into this data set when hard copies of data are digitized. As a result of these and other reasons, uncertainties in the temperature data increase as one goes back in time, particularly given that there are fewer stations early in the record. However, NOAA does not believe these uncertainties are sufficient to undermine the fundamental trends in the data. More information about limitations of early weather data can be found in Kunkel et al. (2005).

3. While heating and cooling degree days provide a good general sense of how temperature changes affect daily life, they are not an exact surrogate for energy use. Many other factors have influenced energy demand over time, such as more energy-efficient heating systems, the introduction and increasingly widespread use of cooling technologies, larger but better-insulated homes, and behavior change. In addition, an indicator of energy use would ideally

account for changes in where people live (e.g., relative population growth in warm regions that require more cooling than heating), which this indicator does not.

10. Sources of Uncertainty

The main source of uncertainty in this indicator relates to the quality of the underlying COOP weather station records. Uncertainty may be introduced into this data set when hard copies of historical data are digitized. As a result of these and other reasons, uncertainties in the temperature data increase as one goes back in time, particularly given that there are fewer stations early in the record. However, NOAA does not believe these uncertainties are sufficient to undermine the fundamental trends in the data. Vose and Menne (2004) suggest that the station density in the U.S. climate network is sufficient to produce robust spatial averages.

NCDC has taken a variety of steps to reduce uncertainties, including correcting the data for time-of-day reporting biases and using the Thom (1954a,b, 1966) methodology to estimate degree days. The value of this approach is that it allows estimation of degree days based on monthly average temperatures, even when the daily data may include some inaccuracies. However, this methodology for estimating HDD and CDD from mean monthly temperatures and the long-term standard deviation of monthly temperatures also introduces some uncertainty. Although this peer-reviewed technique is considered reliable, it could produce inaccurate results if the standard deviation of temperatures has changed over time, for example due to an increasing trend of local variability in daily temperatures.

11. Sources of Variability

HDD and CDD are likely to display the same types of variability as the temperature record on which they are based. Temperatures naturally vary from year to year as a result of normal variation in weather patterns, multi-year climate cycles such as the El Niño–Southern Oscillation and Pacific Decadal Oscillation, and other factors. This indicator accounts for these factors by presenting a long-term record (1895–2013) of how HDD and CDD have changed over time.

12. Statistical/Trend Analysis

To test for the presence of long-term national-level changes in HDD and CDD, the annual average "contiguous 48 states" HDD and CDD data series in Figure 1 were analyzed by ordinary least squares linear regression for the full period of record (1895–2013). Neither trend was statistically significant:

- HDD: regression slope of -0.78 days per year (p = 0.18).
- CDD: regression slope of +0.15 days per year (p = 0.53).

References

Diaz, H.F., and R.G. Quayle. 1980. Heating degree day data applied to residential heating energy consumption. J. Appl. Meteorol. 3:241–246.

Karl, T., C. Williams Jr., P. Young, and W. Wendland. 1986. A model to estimate the time of observation bias associated with monthly mean maximum, minimum, and mean temperatures for the United States. J. Clim. Appl. Meteorol. 25:145–160.

Kunkel, K.E., D.R. Easterling, K. Hubbard, K. Redmond, K. Andsager, M.C. Kruk, and M.L. Spinar. 2005. Quality control of pre-1948 Cooperative Observer Network data. J. Atmos. Ocean. Tech. 22:1691–1705.

NOAA (National Oceanic and Atmospheric Administration). 2002a. Data documentation for data set 9640: Time bias corrected divisional temperature-precipitation-drought index. www1.ncdc.noaa.gov/pub/data/documentlibrary/tddoc/td9640.pdf.

NOAA. 2002b. Data documentation for data set 9641F: Monthly divisional normals and standard deviations of temperature, precipitation, and degree days: 1971–2000 and previous normals periods. www1.ncdc.noaa.gov/pub/data/documentlibrary/tddoc/td9641f.pdf.

NOAA. 2002c. Data documentation for data set 9641G: Monthly and annual heating and cooling degree day normals to selected bases, 1961–90. www1.ncdc.noaa.gov/pub/data/documentlibrary/tddoc/td9641g.pdf.

Thom, H.C.S. 1954a. Normal degree days below any base. Monthly Weather Review 82:111–115.

Thom, H.C.S. 1954b. The rational relationship between heating degree days and temperature. Monthly Weather Review 82:1–6. http://docs.lib.noaa.gov/rescue/mwr/082/mwr-082-01-0001.pdf.

Thom, H.C.S. 1966. Normal degree days above any base by the universal truncation coefficient. Monthly Weather Review 94:461–465.

Vose, R.S., and M.J. Menne. 2004. A method to determine station density requirements for climate observing networks. J. Climate 17(15):2961-2971.

Heat-Related Deaths

Identification

1. Indicator Description

Extreme heat events (i.e., heat waves) have become more frequent in the United States in recent decades (see the High and Low Temperatures indicator), and studies project that the frequency, duration, and intensity of extreme heat events will continue to increase as a consequence of climate change (Melillo et al., 2014). When people are exposed to extreme heat, they can suffer from potentially deadly heat-related illnesses such as heat exhaustion and heat stroke. Thus, as extreme heat events increase, the risk of heat-related deaths and illness is also expected to increase (IPCC, 2014). Tracking the rate of reported heat-related deaths over time provides a key measure of how climate change may affect human well-being.

Components of this indicator include:

- The rate of U.S. deaths between 1979 and 2010 for which heat was classified on death certificates as the underlying (direct) cause (Figure 1, orange line).
- The rate of U.S. deaths between 1999 and 2010 for which heat was classified as either the underlying cause or a contributing factor (Figure 1, blue line).

2. Revision History

April 2010:	Indicator posted.
December 2011:	Updated with data through 2007; added contributing factors analysis to complement the existing time series.
August 2012:	Updated with data through 2009; converted the measure from counts to crude rates; added example figure.
March 2014:	Updated with data through 2010.

Data Sources

3. Data Sources

This indicator is based on data from the U.S. Centers for Disease Control and Prevention's (CDC's) National Vital Statistics System (NVSS), which compiles information from death certificates for nearly every death in the United States. The NVSS is the most comprehensive source of mortality data for the population of the United States. The CDC provided analysis of NVSS data.

Mortality data for the illustrative example figure came from CDC's National Center for Health Statistics (NCHS). The estimate of deaths in excess of the average daily death rate is from the National Research Council's report on climate stabilization targets (NRC, 2011), which cites the peer-reviewed publication Kaiser et al. (2007).

For reference, the illustrative example also shows daily maximum temperature data from the weather station at the Chicago O'Hare International Airport (GHCND:USW00094846).

4. Data Availability

Underlying Causes

The long-term trend line (1979–2010) is based on CDC's Compressed Mortality File, which can be accessed through the CDC WONDER online database at: http://wonder.cdc.gov/mortSQL.html (CDC, 2014a). CDC WONDER provides free public access to mortality statistics, allowing users to query data for the nation as a whole or data broken down by state or region, demographic group (age, sex, race), or International Classification of Diseases (ICD) code. Users can obtain the data for this indicator by accessing CDC WONDER and querying the ICD codes listed in Section 5 for the entire U.S. population.

Underlying and Contributing Causes

The 1999–2010 trend line is based on an analysis developed by the National Environmental Public Health Tracking (EPHT) Program, which CDC coordinates. Monthly totals by state are available online at: http://ephtracking.cdc.gov/showIndicatorPages.action. CDC staff from the National Center for Environmental Health (NCEH) EPHT branch provided national totals to EPA (CDC, 2014b). Users can query underlying and contributing causes of death through CDC WONDER's Multiple Causes of Death file (http://wonder.cdc.gov/mcd.html), but EPHT performed additional steps that cannot be recreated through the publicly available data portal (see Section 6).

Death Certificates

Individual-level data (i.e., individual death certificates) are not publicly available due to confidentiality issues.

Chicago Heat Wave Example

Data for the example figure are based on CDC's Compressed Mortality File, which can be accessed through the CDC WONDER online database at: www.cdc.gov/nchs/data_access/cmf.htm. The analysis was obtained from Kaiser et al. (2007). Daily maximum temperature data for 1995 from the Chicago O'Hare International Airport weather station are available from the National Oceanic and Atmospheric Administration's (NOAA's) National Climatic Data Center (NCDC) at: www.ncdc.noaa.gov/oa/climate/stationlocator.html.

Methodology

5. Data Collection

This indicator is based on causes of death as reported on death certificates. A death certificate typically provides space to designate an immediate cause of death along with up to 20 contributing causes, one of which will be identified as the underlying cause of death. The World Health Organization (WHO) defines the underlying cause of death as "the disease or injury which initiated the train of events leading directly to death, or the circumstances of the accident or violence which produced the fatal injury."

Causes of death are certified by a physician, medical examiner, or coroner, and are classified according to a standard set of codes called the ICD. Deaths for 1979 through 1998 are classified using the Ninth Revision of ICD (ICD-9). Deaths for 1999 and beyond are classified using the Tenth Revision (ICD-10).

Although causes of death rely to some degree on the judgment of the physician, medical examiner, or coroner, the "measurements" for this indicator are expected to be generally reliable based on the medical knowledge required of the "measurer" and the use of a standard classification scheme based on widely accepted scientific definitions. When more than one cause or condition is entered, the underlying cause is determined by the sequence of conditions on the certificate, provisions of the ICD, and associated selection rules and modifications.

Mortality data are collected for the entire population and, therefore, are not subject to sampling design error. For virtually every death that occurs in the United States, a physician, medical examiner, or coroner certifies the causes of death on an official death certificate. State registries collect these death certificates and report causes of death to the NVSS. NVSS's shared relationships, standards, and procedures form the mechanism by which the CDC collects and disseminates the nation's official vital statistics.

Standard forms for the collection of data and model procedures for the uniform registration of death events have been developed and recommended for state use through cooperative activities of the states and CDC's NCHS. All states collect a minimum data set specified by NCHS, including underlying causes of death. CDC has published procedures for collecting vital statistics data (CDC, 1995).

This indicator excludes deaths to foreign residents and deaths to U.S. residents who died abroad.

General information regarding data collection procedures can be found in the Model State Vital Statistics Act and Regulations (CDC, 1995). For additional documentation on the CDC WONDER database (EPA's data source for part of this indicator) and its underlying sources, see: http://wonder.cdc.gov/wonder/help/cmf.html.

CDC has posted a recommended standard certificate of death online at: www.cdc.gov/nchs/data/dvs/DEATH11-03final-ACC.pdf. For a complete list and description of the ICD codes used to classify causes of death, see: www.who.int/classifications/icd/en.

Chicago Heat Wave Example

The mortality data set shown in the example figure includes the entire Standard Metropolitan Statistical Area for Chicago, a region that contains Cook County plus a number of counties in Illinois and Indiana, from June 1 to August 31, 1995.

In the text box above the example figure, values reflect data from Cook County only. The number of deaths classified as "heat-related" on Cook County death certificates between July 11 and July 27, 1995, was reported to CDC by the Cook County Medical Examiner's Office. More information is available in CDC's Morbidity and Mortality Weekly Report (www.cdc.gov/MMWR/preview/mmwrhtml/00038443.htm). Deaths in excess of the average daily death rate for Cook County were determined from death certificates obtained from the Illinois Department of Public Health (Kaiser et al., 2007).

6. Indicator Derivation

This indicator reports the annual rate of deaths per million population that have been classified with ICD codes related to exposure to natural sources of heat. The NVSS collects data on virtually all deaths that occur in the United States, meaning the data collection mechanism already covers the entire target population. Thus, it was not necessary to extrapolate the results on a spatial or population basis. No attempt has been made to reconstruct trends prior to the onset of comprehensive data collection, and no attempt has been made to project data forward into the future.

Underlying Causes

The long-term trend line in Figure 1 reports the rate of deaths per year for which the underlying cause had one of the following ICD codes:

- ICD-9 code E900: "excessive heat—hyperthermia"—specifically subpart E900.0: "due to weather conditions."
- ICD-10 code X30: "exposure to excessive natural heat—hyperthermia."

This component of the indicator is reported for the entire year. EPA developed this analysis based on the publicly available data compiled by CDC WONDER. EPA chose to use crude death rates rather than death counts because rates account for changes in total population over time. Population figures are obtained from CDC WONDER.

Underlying and Contributing Causes

The "underlying and contributing causes" trend line in Figure 1 reports the rate of deaths for which either the underlying cause or the contributing causes had one or more of the following ICD codes:

- ICD-10 code X30: "exposure to excessive natural heat—hyperthermia."
- ICD-10 codes T67.0 through T67.9: "effects of heat and light." Note that the T67 series is used only for contributing causes—never for the underlying cause.

To reduce the chances of including deaths that were incorrectly classified, EPHT did not count the following deaths:

- Deaths occurring during colder months (October through April). Thus, the analysis is limited to May–September.
- Any deaths for which the ICD-10 code W92, "exposure to excessive heat of man-made origin," appears in any cause field. This step removes certain occupational-related deaths.

Foreign residents were excluded. EPHT obtained death counts directly from NVSS, rather than using the processed data available through CDC WONDER. EPHT has not yet applied its methods to data prior to 1999. For a more detailed description of EPHT's analytical methods, see the indicator documentation at: http://ephtracking.cdc.gov/showIndicatorPages.action. Crude death rates were calculated in the same manner as with the underlying causes time series.

The authors of Kaiser et al. (2007) determined that the Chicago area had 692 deaths in excess of the background death rate between June 21 and August 10, 1995. This analysis excluded deaths from accidental causes but included 183 deaths from "mortality displacement," which refers to a decrease in the deaths of individuals who would have died during this period in any case but whose deaths were accelerated by a few days due to the heat wave. This implies that the actual number of excess deaths during the period of the heat wave itself (July 11–27) was higher than 692, but was compensated for by reduced daily death rates in the week after July 27. Thus the value for excess deaths in Cook County for the period of July 11–27 is reported as approximately 700 in the text box above the example figure.

7. Quality Assurance and Quality Control

Vital statistics regulations have been developed to serve as a detailed guide to state and local registration officials who administer the NVSS. These regulations provide specific instructions to protect the integrity and quality of the data collected. This quality assurance information can be found in CDC (1995).

For the "underlying and contributing causes" component of this indicator, extra steps have been taken to remove certain deaths that could potentially reflect a misclassification (see Section 6). These criteria generally excluded only a small number of deaths.

Analysis

8. Comparability Over Time and Space

When plotting the data, EPA inserted a break in the line between 1998 and 1999 to reflect the transition from ICD-9 codes to ICD-10 codes. The change in codes makes it difficult to accurately compare pre-1999 data with data from 1999 and later. Otherwise, all methods have been applied consistently over time and space. ICD codes allow physicians and other medical professionals across the country to use a standard scheme for classifying causes of deaths.

9. Data Limitations

Factors that may impact the confidence, application, or conclusions drawn from this indicator are as follows:

1. It has been well-documented that many deaths associated with extreme heat are not identified as such by the medical examiner and might not be correctly coded on the death certificate. In many cases, they might just classify the cause of death as a cardiovascular or respiratory disease. They might not know for certain whether heat was a contributing factor, particularly if the death did not occur during a well-publicized heat wave. By studying how daily death rates vary with temperature in selected cities, scientists have found that extreme heat contributes to far more deaths than the official death certificates would suggest (Medina-Ramón and Schwartz, 2007). That is because the stress of a hot day can increase the chance of dying from a heart attack, other heart conditions, and respiratory diseases such as pneumonia (Kaiser et al., 2007). These causes of death are much more common than heat-related illnesses such as heat stroke.

Thus, this indicator very likely underestimates the number of deaths caused by exposure to heat. However, it does serve as a reportable national measure of deaths attributable to heat.

2. ICD-9 codes were used to specify underlying cause of death for the years 1979 to 1998. Beginning in 1999, cause of death was specified with ICD-10 codes. The two revisions differ substantially, so data from before 1999 cannot easily be compared with data from 1999 and later.

3. The fact that a death is classified as "heat-related" does not mean that high temperatures were the only factor that caused the death. Pre-existing medical conditions can greatly increase an individual's vulnerability to heat.

4. Heat waves are not the only factor that can affect trends in "heat-related" deaths. Other factors include the vulnerability of the population, the extent to which people have adapted to higher temperatures, the local climate and topography, and the steps people have taken to manage heat emergencies effectively.

5. Heat response measures can make a big difference in death rates. Response measures can include early warning and surveillance systems, air conditioning, health care, public education, infrastructure standards, and air quality management. For example, after a 1995 heat wave, the City of Milwaukee developed a plan for responding to extreme heat conditions in the future. During the 1999 heat wave, this plan cut heat-related deaths nearly in half compared with what was expected (Weisskopf et al., 2002).

10. Sources of Uncertainty

Uncertainty estimates are not available for this indicator. Because statistics have been gathered from virtually the entire target population (i.e., all deaths in a given year), these data are not subject to the same kinds of errors and uncertainties that would be inherent in a probabilistic survey or other type of representative sampling program.

Some uncertainty could be introduced as a result of the professional judgment required of the medical professionals filling out the death certificates, which could potentially result in misclassification or underreporting in some number of cases—probably a small number of cases, but still worth noting.

11. Sources of Variability

There is substantial year-to-year variability within the data, due in part to the influence of a few large events. Many of the spikes apparent in Figure 1 can be attributed to specific severe heat waves occurring in large urban areas.

12. Statistical/Trend Analysis

This indicator does not report on the slope of the apparent trends in heat-related deaths, nor does it calculate the statistical significance of these trends.

References

CDC (U.S. Centers for Disease Control and Prevention). 1995. Model State Vital Statistics Act and Regulations (revised April 1995). DHHS publication no. (PHS) 95-1115. www.cdc.gov/nchs/data/misc/mvsact92aacc.pdf.

CDC. 2014a. CDC Wide-ranging Online Data for Epidemiologic Research (WONDER). Compressed mortality file, underlying cause of death. 1999–2010 (with ICD-10 codes) and 1979–1998 (with ICD-9 codes). Accessed March 2014. http://wonder.cdc.gov/mortSQL.html.

CDC. 2014b. Indicator: Heat-related mortality. National Center for Health Statistics. Annual national totals provided by National Center for Environmental Health staff in March 2014. http://ephtracking.cdc.gov/showIndicatorPages.action.

IPCC (Intergovernmental Panel on Climate Change). 2014. Climate change 2014: Impacts, adaptation, and vulnerability. Working Group II contribution to the IPCC Fifth Assessment Report. Cambridge, United Kingdom: Cambridge University Press.

Kaiser, R., A. Le Tertre, J. Schwartz, C.A. Gotway, W.R. Daley, and C.H. Rubin. 2007. The effect of the 1995 heat wave in Chicago on all-cause and cause-specific mortality. Am. J. Public Health 97(Supplement 1):S158–S162.

Medina-Ramón, M., and J. Schwartz. 2007. Temperature, temperature extremes, and mortality: A study of acclimatization and effect modification in 50 U.S. cities. Occup. Environ. Med. 64(12):827–833.

Melillo, J.M., T.C. Richmond, and G.W. Yohe (eds.). 2014. Climate change impacts in the United States: The third National Climate Assessment. U.S. Global Change Research Program. http://nca2014.globalchange.gov.

NRC (National Research Council). 2011. Climate stabilization targets: Emissions, concentrations, and impacts over decades to millennia. Washington, DC: National Academies Press.

Weisskopf, M.G., H.A. Anderson, S. Foldy, L.P. Hanrahan, K. Blair, T.J. Torok, and P.D. Rumm. 2002. Heat wave morbidity and mortality, Milwaukee, Wis, 1999 vs. 1995: An improved response? Am. J. Public Health 92:830–833.

Lyme Disease

Identification

1. Indicator Description

This indicator looks at the incidence of Lyme disease in the United States since 1991. Lyme disease is a tick-borne bacterial illness that can cause fever, fatigue, and joint and nervous system complications. The spread of Lyme disease is affected by tick prevalence; populations and infection rates among host species; human population patterns, awareness, and behavior; habitat; climate; and other factors. Lyme disease may be useful for understanding the long-term effects of climate change on vector-borne diseases because shorter-term variations in weather have less of an impact on ticks than on other disease vectors such as mosquitoes. This is the case for several reasons (Ogden et al., 2013):

- Ticks have a relatively long life cycle, including stages of development that take place in the soil, where temperatures fluctuate less than air temperatures.
- Tick development rates have a delayed response to temperature changes, which minimizes the effects of short-term temperature fluctuations.
- Ticks can take refuge in the soil during periods of extreme heat, cold, drought, or rainfall.
- Ticks are associated with woodland habitats, where microclimates are buffered from temperature extremes that occur in treeless areas.
- Unlike other disease vectors such as mosquitoes, ticks do not have nonparasitic immature feeding stages whose survival is susceptible to short-term changes in weather.

Consequently, in some locations in the United States, Lyme disease incidence would be expected to increase with climate change.

Components of this indicator include:

- Annual incidence of Lyme disease in the United States (Figure 1).
- Change in reported Lyme disease incidence in the Northeast and Upper Midwest (Figure 2).
- Change in incidence and distribution of reported cases of Lyme disease in the Northeast and Upper Midwest (example maps).

2. Revision History

December 2013:   Indicator proposed.

Data Sources

3. Data Sources

This indicator is based on annual numbers of confirmed Lyme disease cases, nationally and by state, compiled by the Centers for Disease Control and Prevention's (CDC) Division of Vector-Borne Diseases

(DVBD). Incidence was calculated using the most recent mid-year population estimates for each year from the U.S. Census Bureau. The example maps also come from CDC.

4. Data Availability

All of the data for this indicator are publicly available on CDC and Census Bureau websites.

EPA obtained the data for this indicator from CDC's website. Prior to 2008, CDC compiled only confirmed cases, but in 2008 they also began to track probable (but unconfirmed) cases. CDC's database allows users to query just the confirmed cases, which EPA used for this indicator.

Although data are available for 1990, this indicator starts in 1991 because Lyme disease did not become an official nationally reportable disease until January 1991. In 1990, some states reported Lyme disease incidence using the standardized case definition that went into effect nationwide in 1991, but other states did not.

CDC's national and state-level data are available online. Through the years, these data have been published in CDC's Morbidity and Mortality Weekly Reports (MMWR), which are available at: www.cdc.gov/mmwr/mmwr_nd/index.html. Data from 2003 onward are also available in tabular form at: www.cdc.gov/lyme/stats/mmwr.html. Underlying county-level data are not available publicly—or they are combined into multi-year averages before being made publicly available—because of concerns about patient confidentiality. Annual maps of reported cases of Lyme disease, as shown in the example figure for this indicator, are posted online at: www.cdc.gov/lyme/stats/index.html.

Following CDC's standard practice, incidence has been calculated using population estimates on July 1 of each calendar year. These population estimates are publicly available from the U.S. Census Bureau's Population Estimates Program. Pre-2010 data are available at: www.census.gov/popest/data/intercensal/index.html. Data for 2010 and later are available at: www.census.gov/popest/data/index.html.

Methodology

5. Data Collection

This indicator is based on the annual reported number of Lyme disease cases as compiled by CDC.

State and local health departments report weekly case counts for Lyme disease following CDC's case definitions through the National Notifiable Diseases Surveillance System (NNDSS). The NNDSS is a public health system for the reporting of individual cases of disease and conditions to state, local, and territorial health departments, which then forward case information to CDC. The provisional state-level data are reported in CDC's MMWR. After all states have verified their data, CDC publishes an annual surveillance summary for Lyme disease and other notifiable diseases.

Health care providers nationwide follow a standardized definition for what constitutes a "confirmed" case of Lyme disease, but this definition has changed over time (see Section 8). The first standardized surveillance case definition was established in 1990 by the Council of State and Territorial Epidemiologists (CSTE). In January 1991, Lyme disease became a nationally notifiable disease in the

United States, using the CSTE's 1990 definition. As such, state and local health departments work with health care providers to obtain case reports for Lyme disease based upon the CSTE case definition.

6. Indicator Derivation

Figure 1. Reported Cases of Lyme Disease in the United States, 1991–2012

National incidence of Lyme disease was calculated using the total number of confirmed Lyme disease cases and the national population for each year from 1991 through 2012. EPA calculated incidence by dividing the number of confirmed cases per year by the corresponding population on July 1 in the same calendar year. EPA then multiplied the per-person rate by 100,000 to generate a normalized incidence rate per 100,000 people. This is CDC's standard method of expressing the incidence of Lyme disease.

Figure 2. Change in Reported Lyme Disease Incidence in the Northeast and Upper Midwest, 1991–2012

EPA used ordinary least-squares linear regression to determine the slope of the trend over time for each state. Of the 50 states plus the District of Columbia, 32 have a long-term linear trend in Lyme disease incidence that is statistically significant to a 95 percent level, and 31 have trends that are significant to a 99 percent level. However, many of these trends have a very small slope. Taking the regression slope (the annual rate of change) and multiplying it by 21 years (the length of the period of record) to estimate total change, more than half of the states had a total change of less than 1 case per 100,000 in either direction.

In this analysis, 14 states stand out because they have Lyme disease rates more than 10 times higher than most of the other states, average rates of more than 10 cases per 100,000 per year during the most recent five years of data, and in all but three of these states, statistically significant total increases of 10 to 100 cases per 100,000 between 1991 and 2012. These 14 states are:

- Connecticut
- Delaware
- Maine
- Maryland
- Massachusetts
- Minnesota
- New Hampshire
- New Jersey
- New York
- Pennsylvania
- Rhode Island
- Vermont
- Virginia
- Wisconsin

Together, these 14 states accounted for about 95 percent of the nation's reported cases of Lyme disease in 2012, as the map in Figure TD-1 indicates.

Figure TD-1. Reported Cases of Lyme Disease in the United States, 2012

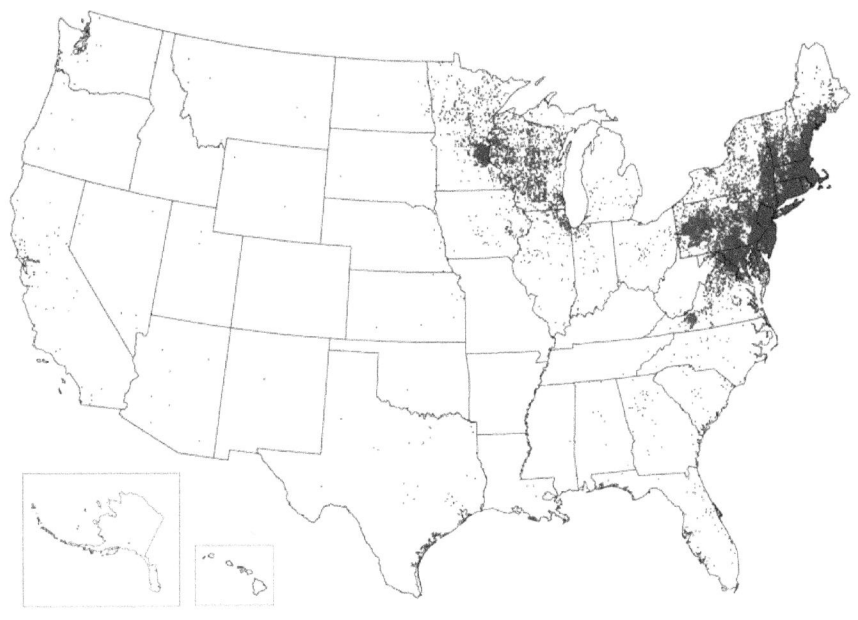

1 dot placed randomly within county of residence for each confirmed case

Data source: CDC: www.cdc.gov/lyme/stats/maps/map2012.html. Accessed April 2014.

Figure 2 shows the total change (annual rate of change [regression slope] multiplied by 21 years) for the 14 states listed above. Trends are not shown for Connecticut, New York, and Rhode Island in Figure 2 because of too much year-to-year variation in reporting practices to allow trend calculation (see Section 12).

Example: Reported Lyme Disease Cases in 1996 and 2012

This example presents two maps—one for the year 1996 and one for the year 2012—to illustrate changes in the incidence and distribution of reported cases of Lyme disease in the United States over time. CDC created these maps. Each dot on the maps represents an individual case placed randomly within the patient's county of residence, which may differ from the county of exposure.

Indicator Development

In the course of developing and revising this indicator based on peer review and comments from CDC experts, EPA considered several ways to present the data. For example:

- The incidence of a disease can be tracked with total case counts or with incidence rates that are normalized by population size. EPA chose to display rates for this indicator so as to eliminate state-to-state population differences and changes in population over time as confounding factors. This approach is also consistent with data for EPA's Heat-Related Deaths indicator, which is displayed using incidence rates.

- EPA considered focusing the analysis of reported Lyme disease on a subset of states. One approach was to consider "reference states" as defined by CDC (www.cdc.gov/mmwr/pdf/ss/ss5710.pdf). However, upon clarification from CDC, this set of reference states has not been used operationally since CDC's Healthy People 2010 effort, which concluded in 2010, and they do not necessarily represent a consistent baseline from which to track trends. EPA chose to use more objective, data-driven thresholds for selecting states to show readers all change in reported Lyme disease incidence as in Figure 2. However, there is scientific evidence (e.g., Diuk-Wasser et al., 2012; Stromdahl and Hickling, 2012) that notes the geographic differences in *Ixodes scapularis* (the deer tick or blacklegged tick) in North America— and that increases in Lyme disease cases in many states south of 35°N latitude are likely due to non-climate-related expansion of northern *I. scapularis* tick genotypes. Analyzing data for a set of states in the northern part of the range of *I. scapularis* might lead to better understanding of changes in Lyme disease cases as related to a warming climate. Thus, future work on this indicator will attempt to reflect the effects of climate change on expansion in the range of *I. scapularis*, increasing abundance of *I. scapularis* where it already occurs, increases in the prevalence of *Borrelia burgdorferi* (the bacteria that actually cause Lyme disease) in host-seeking ticks, and/or updated understanding of other known environmental drivers, such as deer density and changes in landscape, habitat, and biodiversity.

- EPA considered mapping rates or trends by county. However, county-level case totals are only publicly available from CDC in five-year bins, in part because of the very low number of cases reported in many counties.

7.   Quality Assurance and Quality Control

Each state has established laws mandating that health providers report cases of various diseases (including Lyme disease) to their health departments. Each state health department verifies its data before sharing them with CDC. The NNDSS is the primary system by which health surveillance data are conveyed to CDC for national-level analyses.

Starting in 1990, CDC launched the National Electronic Telecommunications System for Surveillance (NETSS), replacing mail and phone-based reporting. In 2000, CDC developed the National Electronic Disease Surveillance System (NEDSS) Base System (NBS). This central reporting system sets data and information technology standards for departments that provide data to CDC, ensuring data are submitted quickly, securely, and in a consistent format.

Using CSTE case definitions, CDC provides state and local health departments and health providers with comprehensive guidance on laboratory diagnosis and case classification criteria, ensuring that all health providers and departments classify Lyme disease cases consistently throughout the United States.

State health officials use various methods to ascertain cases, including passive surveillance initiated by health care providers, laboratory-based surveillance, and "enhanced or active surveillance" (Bacon et al., 2008). State officials check the data and remove duplicate reports before submitting annual totals to CDC.

CDC has undertaken a review of alternative data sources to see how closely they align with the disease counts captured by the NNDSS. These alternative sources include medical claims information from a large insurance database, a survey of clinical laboratories, and a survey that asks individuals whether they have been diagnosed with Lyme disease in the previous year. Preliminary results from this review suggest that the NNDSS may be undercounting the true number of cases of Lyme disease (CDC, 2013). See Section 10 for further discussion about this possible source of uncertainty.

Analysis

8. Comparability Over Time and Space

Lyme disease data collection follows CDC's case definition to ensure consistency and comparability across the country. However, the national case definition for Lyme disease has changed twice since Lyme disease became a notifiable disease: first in 1996 and again in 2008. Prior to 1996, a confirmed case of Lyme disease required only a skin lesion with the characteristic "bulls-eye" appearance. In 1996, CDC expanded the definition of confirmed cases to include laboratory-confirmed, late manifestation symptoms such as issues with the musculoskeletal, nervous, and cardiovascular systems. In 2008, the case classifications were expanded again to include suspected and probable cases.

These definition changes necessitate careful comparisons of data from multiple years. While it is not possible to control for the case definition change in 1996, CDC provides the numbers of confirmed cases and suspected and probable cases separately. The granularity of the data enables EPA to use confirmed cases in the incidence rate calculation for all years and exclude the probable cases that have been counted since 2008, ensuring comparability over time.

In addition to the national changes, several state reporting agencies have changed their own definitions at various times. These state-level changes include California in 2005, Connecticut in 2003, the District of Columbia in 2011, Hawaii in 2006, New York in 2007, and Rhode Island in 2004. The extent to which these changes affect overall trends is unknown, but it is worth noting that Connecticut and Rhode Island both have apparent discontinuities in their annual totals around the time of their respective definitional changes, and these two states and New York all have statistically insignificant long-term trends (see Section 12), despite being surrounded by states with statistically significant increases. Because of these state-level uncertainties, Figure 2 shows only state-level trends that are statistically significant. In this case, the p-value for each displayed state is less than 0.001.

9. Data Limitations

Factors that may have an impact on the confidence, application, or conclusions drawn from this indicator are as follows:

1. For consistency, this indicator includes data for only confirmed cases of Lyme disease. However, changes in diagnosing practices and awareness of the disease over time can affect trends.

2. CDC's national Lyme disease case definitions have changed twice since Lyme disease became a notifiable disease. As discussed in Section 8, it is not possible to control for the case definition change in 1996, which adds some uncertainty to the indicator. State agencies have also changed their definitions at various times, as described in Section 8.

3. As described in Section 10, public health experts believe that many cases of Lyme disease are not reported, which means this indicator underestimates the true incidence of the disease (CDC,

2013). The reporting rate may vary over time and space as a result of differences in funding and emphasis among state surveillance programs. In addition, cases of Lyme disease in locations where Lyme disease is not endemic may be unidentified or misdiagnosed.

4. As an indicator of climate change, Lyme disease is limited due to several confounding factors:

- Pest extermination efforts and public health education may counteract the growth of confirmed cases expected due to warming climates.

- Importantly, there are several factors driving changes in incidence of Lyme disease other than climate. Several of these factors have not been well-quantified or studied. Possible factors include range expansion of vector ticks, which is not always climate-related; proximity of hosts; changes in deer density; changes in biodiversity; and the effects of landscape changes such as suburbanization, deforestation, and reforestation.

- Pathogen transmission is affected by several factors including geographic distribution, population density, prevalence of infection by zoonotic pathogens, and the pathogen load within individual hosts and vectors (e.g., Cortinas and Kitron, 2006; Lingren, 2005; Mills et al., 2010; Raizman, 2013).

- Human exposure depends upon socioeconomic and cultural factors, land use, health care access, and living conditions (Gage et al., 2008; Gubler et al., 2001; Hess et al., 2012; Lafferty, 2009; Wilson, 2009).

5. Lyme disease surveillance data capture the county of residence, which is not necessarily the location where an individual was infected.

10. Sources of Uncertainty

The main source of uncertainty for this indicator stems from its dependence on surveillance data. Surveillance data can be subject to underreporting and misclassification. Because Lyme disease is often determined based upon clinical symptoms, lack of symptoms or delayed symptoms may result in overlooked or misclassified cases. Furthermore, surveillance capabilities can vary from state to state, or even from year to year based upon budgeting and personnel.

Although Lyme disease cases are supposed to be reported to the NNDSS, reporting is actually voluntary. As a result, surveillance data for Lyme disease do not provide a comprehensive determination of the U.S. population with Lyme disease. For example, it has been reported that the annual total number of people diagnosed with Lyme disease may be as much as 10 times higher than the surveillance data indicate (CDC, 2013). Consequently, this indicator provides an illustration of trends over time, not a measure of the exact number of Lyme disease cases in the United States.

Another issue is that surveillance data are captured by county of residence rather than county of exposure. Reports of Lyme disease may therefore occur in states with no active pathogen populations. For example, a tourist may be infected with Lyme disease while visiting Connecticut (an area with high incidence of Lyme disease) but not be identified as a Lyme disease case until the tourist returns home to Florida (an area where blacklegged ticks cannot survive). This may result in underreporting in areas of high Lyme disease incidence and overreporting in areas of low Lyme disease incidence.

For a discussion of the uncertainties associated with the U.S. Census Bureau's intercensal estimates, see: www.census.gov/popest/methodology/intercensal_nat_meth.pdf.

11. Sources of Variability

The incidence of Lyme disease is likely to display variability over time and space due to:

- Changes in populations of blacklegged ticks and host species (e.g., deer, mice, birds) over time.
- Spatial distribution of blacklegged ticks and changes in their distribution over time.
- The influence of climate on the activity and seasonality of the blacklegged tick.
- Variability in human population over time and space.

This indicator accounts for these factors by presenting a broad multi-decadal national trend in Figures 1 and 2. EPA has reviewed the statistical significance of these trends (see Section 12).

12. Statistical/Trend Analysis

Based on ordinary least-squares linear regression, the national incidence rate in Figure 1 increases at an average annual rate of +0.24 cases per 100,000 people ($p < 0.001$).

Of the 14 states shaded in Figure 2, 11 had statistically significant increases in their annual incidence rates from 1991 to 2012 (all p-values < 0.001), based on ordinary least-squares linear regression. The shading in Figure 2 shows the magnitude of these trends. The other three states did not: Connecticut ($p = 0.96$), New York ($p = 0.34$), and Rhode Island ($p = 0.07$). A broader analysis described in Section 6 found that more than half of the 50 states had significant trends in their annual incidence rates from 1991 to 2012, but most of these states were excluded from Figure 2 because their overall incidence rates have consistently been at least an order of magnitude lower than the rates in the 14 key Northeast and Upper Midwest states where Lyme disease is most prevalent.

References

Bacon, R.M., K.J. Kugeler, and P.S. Mead. 2008. Surveillance for Lyme disease—United States, 1992–2006. Morbidity and Mortality Weekly Report 57(SS10):1–9.

CDC (U.S. Centers for Disease Control and Prevention). 2013. CDC provides estimate of Americans diagnosed with Lyme disease each year. www.cdc.gov/media/releases/2013/p0819-lyme-disease.html.

Cortinas, M.R., and U. Kitron. 2006. County-level surveillance of white-tailed deer infestation by *Ixodes scapularis* and *Dermacentor albipictus* (Acari: Ixodidae) along the Illinois River. J. Med. Entomol. 43(5):810–819.

Diuk-Wasser, M.A., A.G. Hoen, P. Cislo, R. Brinkerhoff, S.A. Hamer, M. Rowland, R. Cortinas, G. Vourc'h, F. Melton, G.J. Hickling, J.I. Tsao, J. Bunikis, A.G. Barbour, U. Kitron, J. Piesman, and D. Fish. 2012. Human risk of infection with *Borrelia burgdorferi*, the Lyme disease agent, in eastern United States. Am. J. Trop. Med. Hyg. 86(2):320–327.

Gage, K.L., T.R. Burkot, R.J. Eisen, and E.B. Hayes. 2008. Climate and vector-borne diseases. A. J. Prev. Med. 35(5):436–450.

Gubler, D.J., P. Reiter, K.L. Ebi, W. Rap, R. Nasci, and J.A. Patz. 2001. Climate variability and change in the United States: Potential impacts on vector- and rodent-borne diseases. Environ. Health. Perspect. 109:223–233.

Hess, J.J., J.Z. McDowell, and G. Luber. 2012. Integrating climate change adaptation into public health practice: Using adaptive management to increase adaptive capacity and build resilience. Environ. Health. Perspect. 120(2):171–179.

Lafferty, K.D. 2009. The ecology of climate change and infectious diseases. Ecology 90(4):888–900.

Lingren, M., W.A. Rowley, C. Thompson, and M. Gilchrist. 2005. Geographic distribution of ticks (Acari: Ixodidae) in Iowa with emphasis on *Ixodes scapularis* and their infection with *Borrelia burgdorferi*. Vector-Borne Zoonot. 5(3):219–226.

Mills, J.N., K.L. Gage, and A.S. Khan. 2010. Potential influence of climate change on vector-borne and zoonotic diseases: A review and proposed research plan. Environ. Health. Perspect. 118(11):1507–1514.

Ogden, N.H., S. Mechai, and G. Margos. 2013. Changing geographic ranges of ticks and tick-borne pathogens: Drivers, mechanisms, and consequences for pathogen diversity. Front. Cell. Infect. Microbiol. 3:46.

Raizman, E.A., J.D. Holland, and J.T. Shukle. 2013. White-tailed deer (*Odocoileus virginianus*) as a potential sentinel for human Lyme disease in Indiana. Zoonoses Public Hlth. 60(3):227–233.

Stromdahl, E.Y., and G.J. Hickling. 2012. Beyond Lyme: Aetiology of tick-borne human diseases with emphasis on the south-eastern United States. Zoonoses Public Hlth. 59 Suppl 2:48–64.

Wilson, K. 2009. Climate change and the spread of infectious ideas. Ecology 90:901–902.

Length of Growing Season

Identification

1. Indicator Description

This indicator measures the length of the growing season (or frost-free season) in the contiguous 48 states between 1895 and 2013. The growing season often determines which crops can be grown in an area, as some crops require long growing seasons, while others mature rapidly. Growing season length is limited by many different factors. Depending on the region and the climate, the growing season is influenced by air temperatures, frost days, rainfall, or daylight hours. Air temperatures, frost days, and rainfall are all associated with climate, so these drivers of the growing season could change as a result of climate change.

This indicator focuses on the length of the growing season as defined by frost-free days. Components of this indicator include:

- Length of growing season in the contiguous 48 states, both nationally (Figure 1) and for the eastern and western halves of the country (Figure 2).
- Timing of the last spring frost and the first fall frost in the contiguous 48 states (Figure 3).

2. Revision History

April 2010:	Indicator posted.
December 2011:	Updated with data through 2010.
April 2012:	Updated with data through 2011.
August 2013:	Updated on EPA's website with data through 2012.
March 2014:	Updated with data through 2013.

Data Sources

3. Data Sources

Data were provided by Dr. Kenneth Kunkel of the National Oceanic and Atmospheric Administration's (NOAA's) Cooperative Institute for Climate and Satellites (CICS), who analyzed minimum daily temperature records from weather stations throughout the contiguous 48 states. Temperature measurements come from weather stations in NOAA's Cooperative Observer Program (COOP).

4. Data Availability

EPA obtained the data for this indicator from Dr. Kenneth Kunkel at NOAA CICS. Dr. Kunkel had published an earlier version of this analysis in the peer-reviewed literature (Kunkel et al., 2004), and he provided EPA with an updated file containing growing season data through 2013.

All raw COOP data are maintained by NOAA's National Climatic Data Center (NCDC). Complete COOP data, embedded definitions, and data descriptions can be downloaded from the Web at: www.ncdc.noaa.gov/doclib. State-specific data can be found at: www7.ncdc.noaa.gov/IPS/coop/coop.html;jsessionid=312EC0892FFC2FBB78F63D0E3ACF6CBC. There are no confidentiality issues that could limit accessibility, but some portions of the data set might need to be formally requested. Complete metadata for the COOP data set can be found at: www.nws.noaa.gov/om/coop.

Methodology

5. Data Collection

This indicator focuses on the timing of frosts, specifically the last frost in spring and the first frost in fall. It was developed by analyzing minimum daily temperature records from COOP weather stations throughout the contiguous 48 states.

COOP stations generally measure temperature at least hourly, and they record the minimum temperature for each 24-hour time span. Cooperative observers include state universities, state and federal agencies, and private individuals whose stations are managed and maintained by NOAA's National Weather Service (NWS). Observers are trained to collect data, and the NWS provides and maintains standard equipment to gather these data. The COOP data set represents the core climate network of the United States (Kunkel et al., 2005). Data collected by COOP sites are referred to as U.S. Daily Surface Data or Summary of the Day data.

The study on which this indicator is based includes data from 750 stations in the contiguous 48 states. These stations were selected because they met criteria for data availability; each station had to have less than 10 percent of temperature data missing over the period from 1895 to 2013. For a map of these station locations, see Kunkel et al. (2004). Pre-1948 COOP data were previously only available in hard copy, but were recently digitized by NCDC, thus allowing analysis of more than 100 years of weather and climate data.

Temperature monitoring procedures are described in the full metadata for the COOP data set available at: www.nws.noaa.gov/om/coop. General information on COOP weather data can be found at: www.nws.noaa.gov/os/coop/what-is-coop.html.

6. Indicator Derivation

For this indicator, the length of the growing season is defined as the period of time between the last frost of spring and the first frost of fall, when the air temperature drops below the freezing point of 32°F. Minimum daily temperature data from the COOP data set were used to determine the dates of last spring frost and first fall frost using an inclusive threshold of 32°F. Methods for producing regional and national trends were designed to weight all regions evenly regardless of station density.

Figure 1 shows trends in the overall length of the growing season, which is the number of days between the last spring frost and the first fall frost. Figure 2 shows trends in the length of growing season for the eastern United States versus the western United States, using 100°W longitude as the dividing line between the two halves of the country. Figure 3 shows trends in the timing of the last spring frost and the first fall frost, also using units of days.

All three figures show the deviation from the 1895–2013 long-term average, which is set at zero for reference. Thus, if spring frost timing in year *n* is shown as -4, it means the last spring frost arrived four days earlier than usual. Note that the choice of baseline period will not affect the shape or the statistical significance of the overall trend; it merely moves the trend up or down on the graph in relation to the point defined as "zero."

To smooth out some of the year-to-year variability and make the results easier to understand visually, all three figures plot 11-year moving averages rather than annual data. EPA chose this averaging period to be consistent with the recommended averaging method used by Kunkel et al. (2004) in an earlier version of this analysis. Each average is plotted at the center of the corresponding 11-year window. For example, the average from 2003 to 2013 is plotted at year 2008. EPA used endpoint padding to extend the 11-year smoothed lines all the way to the ends of the period of record. Per the data provider's recommendation, EPA calculated smoothed values centered at 2009, 2010, 2011, 2012, and 2013 by inserting the 2008–2013 average into the equation in place of the as-yet unreported annual data points for 2014 and beyond. EPA used an equivalent approach at the beginning of the time series.

Kunkel et al. (2004) provide a complete description of the analytical procedures used to determine length of growing season trends. No attempt has been made to represent data outside the contiguous 48 states or to estimate trends before or after the 1895–2013 time period.

7. Quality Assurance and Quality Control

NOAA follows extensive quality assurance and quality control (QA/QC) procedures for collecting and compiling COOP weather station data. For documentation of COOP methods, including training manuals and maintenance of equipment, see: www.nws.noaa.gov/os/coop/training.htm. These training materials also discuss QC of the underlying data set. Pre-1948 COOP data were recently digitized from hard copy. Kunkel et al. (2005) discuss QC steps associated with digitization and other factors that might introduce error into the growing season analysis.

The data used in this indicator were carefully analyzed in order to identify and eliminate outlying observations. A value was identified as an outlier if a climatologist judged the value to be physically impossible based on the surrounding values, or if the value of a data point was more than five standard deviations from the station's monthly mean. Readers can find more details on QC analysis for this indicator in Kunkel et al. (2004) and Kunkel et al. (2005).

Analysis

8. Comparability Over Time and Space

Data from individual weather stations were averaged in order to determine national and regional trends in the length of growing season and the timing of spring and fall frosts. To ensure spatial balance, national and regional values were computed using a spatially weighted average, and as a result, stations in low-station-density areas make a larger contribution to the national or regional average than stations in high-density areas.

9. Data Limitations

Factors that may impact the confidence, application, or conclusions drawn from this indicator are as follows:

1. Changes in measurement techniques and instruments over time can affect trends. However, these data were carefully reviewed for quality, and values that appeared invalid were not included in the indicator. This indicator includes only data from weather stations that did not have many missing data points.

2. The urban heat island effect can influence growing season data; however, these data were carefully quality-controlled and outlying data points were not included in the calculation of trends.

10. Sources of Uncertainty

Kunkel et al. (2004) present uncertainty measurements for an earlier (but mostly similar) version of this analysis. To test worst-case conditions, Kunkel et al. (2004) computed growing season trends for a thinned-out subset of stations across the country, attempting to simulate the density of the portions of the country with the lowest overall station density. The 95 percent confidence intervals for the resulting trend in length of growing season were ±2 days. Thus, there is very high likelihood that observed changes in growing season are real and not an artifact of sampling error.

11. Sources of Variability

At any given location, the timing of spring and fall frosts naturally varies from year to year as a result of normal variation in weather patterns, multi-year climate cycles such as the El Niño–Southern Oscillation and Pacific Decadal Oscillation, and other factors. This indicator accounts for these factors by applying an 11-year smoothing filter and by presenting a long-term record (more than a century of data). Overall, variability should not impact the conclusions that can be inferred from the trends shown in this indicator.

12. Statistical/Trend Analysis

EPA calculated long-term trends by ordinary least-squares regression to support statements in the "Key Points" text. Kunkel et al. (2004) determined that the overall increase in growing season was statistically significant at a 95 percent confidence level in both the East and the West.

References

Kunkel, K.E., D.R. Easterling, K. Hubbard, and K. Redmond. 2004. Temporal variations in frost-free season in the United States: 1895–2000. Geophys. Res. Lett. 31:L03201.

Kunkel, K.E., D.R. Easterling, K. Hubbard, K. Redmond, K. Andsager, M.C. Kruk, and M.L. Spinar. 2005. Quality control of pre-1948 Cooperative Observer Network data. J. Atmos. Ocean. Tech. 22:1691–1705.

Ragweed Pollen Season

Identification

1. Indicator Description

This indicator describes trends in the annual length of pollen season for ragweed (*Ambrosia* species) at 11 North American sites from 1995 to 2013. In general, by leading to more frost-free days and warmer seasonal air temperatures, climate change can contribute to shifts in flowering time and pollen initiation from allergenic plant species, and increased carbon dioxide concentrations alone can elevate the production of plant-based allergens (Melillo et al., 2014). In the case of ragweed, the pollen season begins with the shift to shorter daylight after the summer solstice, and it ends in response to cold weather in the fall (i.e., first frost). These constraints suggest that the length of ragweed pollen season is sensitive to climate change by way of changes to fall temperatures. Because allergies are a major public health concern, observed changes in the length of the ragweed pollen season over time provide insight into ways in which climate change may affect human well-being.

2. Revision History

December 2011:	Indicator developed.
May 2012:	Updated with data through 2011.
May 2014:	Updated with data through 2013.

Data Sources

3. Data Sources

Data for this indicator come from the National Allergy Bureau. As a part of the American Academy of Allergy, Asthma, and Immunology's (AAAAI's) Aeroallergen Network, the National Allergy Bureau collects pollen data from dozens of stations around the United States. Canadian pollen data originate from Aerobiology Research Laboratories. The data were compiled and analyzed for this indicator by a team of researchers who published a more detailed version of this analysis in 2011, based on data through 2009 (Ziska et al., 2011).

4. Data Availability

EPA acquired data for this indicator from Dr. Lewis Ziska of the U.S. Department of Agriculture, Agricultural Research Service. Dr. Ziska was the lead author of the original analysis published in 2011 (Ziska et al., 2011). He provided an updated version for EPA's indicator, with data through 2013.

Users can access daily ragweed pollen records for each individual U.S. pollen station on the National Allergy Bureau's website at: www.aaaai.org/global/nab-pollen-counts.aspx. *Ambrosia* spp. is classified as a "weed" by the National Allergy Bureau and appears in its records accordingly. Canadian pollen data are not publicly available, but can be purchased from Aerobiology Research Laboratories at: www.aerobiology.ca/products/data.php.

Methodology

5. Data Collection

This indicator is based on daily pollen counts from 11 long-term sampling stations in central North America. Nine sites were in the United States; two sites were in Canada. Sites were selected based on availability of pollen data and nearby weather data (as part of a broader analysis of causal factors) and to represent a variety of latitudes along a roughly north-south transect. Sites were also selected for consistency of elevation and other locational variables that might influence pollen counts.

Table TD-1 identifies the station locations and the years of data available from each station.

Table TD-1. Stations Reporting Ragweed Data for this Indicator

Station (ordered from north to south)	Start year	End year	Notes
Saskatoon, Saskatchewan (Canada)	1994	2013	
Winnipeg, Manitoba (Canada)	1994	2013	
Fargo, North Dakota	1995	2012	Stopped collecting data after 2012
Minneapolis, Minnesota	1991	2013	
La Crosse, Wisconsin	1988	2013	
Madison, Wisconsin	1973	2013	
Papillion/Bellevue, Nebraska	1989	2013	Station was in Papillion until 2012, then moved a few miles away to Bellevue
Kansas City, Missouri	1997	2013	
Rogers, Arkansas	1996	2012	No data available for 2013
Oklahoma City, Oklahoma	1991	2012	No data available for 2013
Georgetown, Texas	1979	2013	Near Austin, Texas

Each station relies on trained individuals to collect air samples. Samples were collected using one of three methods at each counting station:

1. Slide gathering: Blank slides with an adhesive are left exposed to outdoor air to collect airborne samples.

2. Rotation impaction aeroallergen sampler: An automated, motorized device that spins air of a known volume such that airborne particles adhere to a surrounding collection surface.

3. Automated spore sampler from Burkard Scientific: A device that couples a vacuum pump and a sealed rolling tumbler of adhesive paper in a way that records spore samples over time.

Despite differences in sample collection, all sites rely on the human eye to identify and count spores on microscope slides. All of these measurement methods follow standard peer-reviewed protocols. The resulting data sets from AAAAI and Aerobiology Research Laboratories have supported a variety of peer-reviewed studies. Although the sample collection methodologies do not allow for a comparison of total pollen counts across stations that used different methods, the methods are equally sensitive to the appearance of a particular pollen species.

6. Indicator Derivation

By reviewing daily ragweed pollen counts over an entire season, analysts established start and end dates for each location as follows:

- The start date is the point at which 1 percent of the cumulative pollen count for the season has been observed, meaning 99 percent of all ragweed pollen appears after this day.
- The end date is the point at which 99 percent of the cumulative pollen count for the season has been observed.

The duration of pollen season is simply the length of time between the start date and end date.

Two environmental parameters constrain the data used in calculating the length of ragweed season. As a short-day plant, ragweed will not flower before the summer solstice. Furthermore, ragweed is sensitive to frost and will not continue flowering once temperatures dip below freezing (Deen et al., 1998). Because of these two biological constraints, ragweed pollen identified before June 21 or after the first fall frost (based on local weather data) was not included in the analysis.

Once the start date, end date, and total length of the pollen season were determined for each year and location, best-fit regression lines were calculated from all years of available data at each location. Thus, the longer the data record, the more observations that were available to feed into the trend calculation. Next, the regression coefficients were used to define the length of the pollen season at each station in 1995 and 2013. Figure 1 shows the difference between the 2013 season length and the 1995 season length that were calculated using this method.

Ziska et al. (2011) present these analytical methods and describe them in greater detail.

7. Quality Assurance and Quality Control

Pollen counts are determined by trained individuals who follow standard protocols, including procedures for quality assurance and quality control (QA/QC). To be certified as a pollen counter, one must meet various quality standards for sampling and counting proficiency.

Analysis

8. Comparability Over Time and Space

Different stations use different sampling methods, so absolute pollen counts are not comparable across stations. However, because all of the methods are consistent in how they identify the start and end of the pollen season, the season's length data are considered comparable over time and from station to station.

9. Data Limitations

Factors that may impact the confidence, application, or conclusions drawn from this indicator are as follows:

1. This indicator focuses on only 11 stations in the central part of North America. The impacts of climate change on ragweed growth and pollen production could vary in other regions, such as coastal or mountainous areas.

2. This indicator does not describe the extent to which the intensity of ragweed pollen season (i.e., pollen counts) may also be changing.

3. The indicator is sensitive to other factors aside from weather, including the distribution of plant species as well as pests or diseases that impact ragweed or competing species.

4. Although some stations have pollen data dating back to the 1970s, this indicator characterizes trends only from 1995 to 2013, based on data availability for the majority of the stations in the analysis.

10. Sources of Uncertainty

Error bars for the calculated start and end dates for the pollen season at each site were included in the data set that was provided to EPA. Identification of the ragweed pollen season start and end dates may be affected by a number of factors, both human and environmental. For stations using optical identification of ragweed samples, the technicians evaluating the slide samples are subject to human error. Further discussion of error and uncertainty can be found in Ziska et al. (2011).

11. Sources of Variability

Wind and rain may impact the apparent ragweed season length. Consistently windy conditions could keep pollen particles airborne for longer periods of time, thereby extending the apparent season length. Strong winds could also carry ragweed pollen long distances from environments with more favorable growing conditions. In contrast, rainy conditions have a tendency to draw pollen out of the air. Extended periods of rain late in the season could prevent what would otherwise be airborne pollen from being identified and recorded.

12. Statistical/Trend Analysis

The indicator relies on a best-fit regression line for each sampling station to determine the change in ragweed pollen season. Trends in season length over the full period of record were deemed to be statistically significant to a 95 percent level ($p < 0.05$) at five of the 11 stations, based on ordinary least-squares regression: Saskatoon, Saskatchewan; Winnipeg, Manitoba; Minneapolis, Minnesota; La Crosse, Wisconsin; and Madison, Wisconsin. For further discussion and previous significance analysis, see Ziska et al. (2011).

References

Arbes, S.J., Jr., P.J. Gergen, L. Elliott, and D.C. Zeldin. 2005. Prevalences of positive skin test responses to 10 common allergens in the U.S. population: Results from the third National Health and Nutrition Examination Survey. J. Allergy Clin. Immunol. 116(2):377–383.

Deen, W., L.A. Hunt, and C.J. Swanton. 1998. Photothermal time describes common ragweed (*Ambrosia artemisiifolia L.*) phenological development and growth. Weed Sci. 46:561–568.

Melillo, J.M., T.C. Richmond, and G.W. Yohe (eds.). 2014. Climate change impacts in the United States: The third National Climate Assessment. U.S. Global Change Research Program. http://nca2014.globalchange.gov.

Ziska, L., K. Knowlton, C. Rogers, D. Dalan, N. Tierney, M. Elder, W. Filley, J. Shropshire, L.B. Ford, C. Hedberg, P. Fleetwood, K.T. Hovanky, T. Kavanaugh, G. Fulford, R.F. Vrtis, J.A. Patz, J. Portnoy, F. Coates, L. Bielory, and D. Frenz. 2011. Recent warming by latitude associated with increased length of ragweed pollen season in central North America. P. Natl. Acad. Sci. USA 108:4248–4251.

Wildfires

Identification

1. Indicator Description

This indicator tracks wildfire frequency, total burned acreage, and burn severity in the United States from 1983 to 2013. Although wildfires occur naturally and play a long-term role in the health of ecosystems, climate change threatens to increase the frequency, extent, and severity of fires through increased temperatures and drought. Earlier spring melting and reduced snowpack result in decreased water availability during hot summer conditions, which in turn contributes to an increased risk of wildfires, allowing fires to start more easily and burn hotter. Thus, while climate change is not the only factor that influences patterns in wildfire, the many connections between wildfire and climate make this indicator a useful tool for examining a possible impact of climate change on ecosystems and human well-being. Wildfires are also relevant to climate because they release carbon dioxide into the atmosphere, which in turn contributes to additional climate change.

Components of this indicator include:

- Wildfire frequency (Figure 1).
- Burned acreage from wildfires (Figure 2).
- Wildfire burn severity (Figure 3).
- Burned acreage from wildfires by state over time (Figure 4).

2. Revision History

December 2013:   Indicator proposed.
April 2014:   Figures 1 and 2 updated with data through 2013; Figures 3 and 4 updated with
                   data through 2012.

Data Sources

3. Data Sources

Wildfire data come from three sources:

1. Summary data for wildfire frequency and burned acreage from 1983 through 2013 (Figures 1 and 2) are provided by the National Interagency Coordination Center (NICC), housed within the National Interagency Fire Center (NIFC).

2. For comparison in Figures 1 and 2, EPA obtained a data set called the United States Department of Agriculture (USDA) Forest Service Wildfire Statistics, which provides annual frequency and burned acreage totals through 1997 based on a different counting approach.

3. Burn severity (Figure 3) and state-by-state burn acreage (Figure 4) data were obtained from the Monitoring Trends in Burn Severity (MTBS) project, sponsored by the Wildland Fire Leadership Council (WFLC). The MTBS is a joint project of the USDA Forest Service Remote Sensing Applications Center (RSAC) and the United States Geological Survey (USGS) Earth Resources Observation and Science (EROS) Center. Other collaborators include the National Park Service, other USGS and USDA research facilities, and various academic institutions. The project provides data on individual wildfire incidents that meet certain size criteria (≥ 1,000 acres in the western United States or ≥ 500 acres in the eastern United States). These data were available from 1984 to 2012.

The analysis in Figure 4 normalizes wildfire extent by the land area of each state. Land areas come from the U.S. Census Bureau.

4. Data Availability

NIFC data for annual trends in wildfire frequency and acreage are available from the NIFC website at: www.nifc.gov/fireInfo/fireInfo_statistics.html. These NIFC data are also mirrored in the annual Wildland Fire Summary and Statistics reports from 2000 through 2013 at: www.predictiveservices.nifc.gov/intelligence/intelligence.htm. NIFC totals are based on raw fire incidence data reported via the Incident Command System (ICS) Incident Status Summary Reports (ICS-209 forms). Some raw ICS-209 forms are available for individual viewing at: http://fam.nwcg.gov/fam-web/hist_209/report_list_209.

The USDA Forest Service Wildfire Statistics represent a complementary approach to compiling fire occurrence and extent data. These statistics come from annual Forest Service reports officially known as annual "Wildland Fire Statistics," but more commonly called "Smokey Bear Reports." These compilation reports are based on reports submitted to the Forest Service by individual state and federal agencies, covering land within each agency's jurisdiction. Smokey Bear Reports were provided to EPA by Forest Service staff.

MTBS project analyses use raw ICS-209 form data from 1984 to 2012 as the basis for further processing. Summary data are publicly available at: http://mtbs.gov/dataaccess.html. This online database search tool also provides more detailed and comprehensive records, including burned area classification for each individual fire incident. Detailed records for this indicator were provided by MTBS staff.

The U.S. Census Bureau has published official land areas for each state in the Statistical Abstract of the United States, available online at: www.census.gov/compendia/statab/2012/tables/12s0358.pdf.

Methodology

5. Data Collection

This indicator presents three measures of wildfires over time reported on an annual basis: (1) the total number of wildfires, (2) acreage burned by wildfires, and (3) the burn severity of those fires. For the purposes of this indicator, wildfires encompass "unplanned, unwanted wildland fire[s] including unauthorized human-caused fires, escaped wildland fire use events, escaped prescribed fire projects, and all other wildland fires where the objective is to put the fire out" (NWCG, 2012). A wildland is defined as "an area in which development is essentially non-existent, except for roads, railroads, powerlines, and similar transportation facilities." Fire severity is defined as the "degree to which a site has been altered or disrupted by fire; loosely a product of fire intensity and residence time." These data cover all fifty states.

Figures 1 and 2. Wildfire Frequency and Acreage in the United States, 1983–2013

Wildfire frequency and burn acreage data are based upon local-, state-, and national-level reporting of wildland fire incidents submitted to the NIFC via the ICS-209 form (Fire and Aviation Management and Predictive Services, 2009). The data captured in these forms can also be submitted to the NIFC using the Incident Management Situation (SIT)-209 reporting application. The ICS-209 guidelines require that large fires (100+ acres in timber and 300+ acres in grasslands) must be reported, but they do not set a minimum fire size for reporting. Thus, the data set includes small fires, including some that may have burned just a few acres or less.

Supplementary data come from annual Smokey Bear Reports, which are based on annual reports submitted to the Forest Service by individual state and federal agencies. These original reports describe fires taking place on land within each reporting agency's fire protection jurisdiction. The USDA Forest Service stopped compiling Smokey Bear Reports after 1997.

Figure 3. Damage Caused by Wildfires in the United States, 1984–2012

MTBS uses satellite imagery to map burn severity and perimeters of large fires (≥ 1,000 acres in the western United States or ≥ 500 acres in the eastern United States). These thresholds are applied based on the "West" and "East" regions shown in Figure TD-1.

Figure TD-1. Region Boundaries for MTBS Size Threshold Application

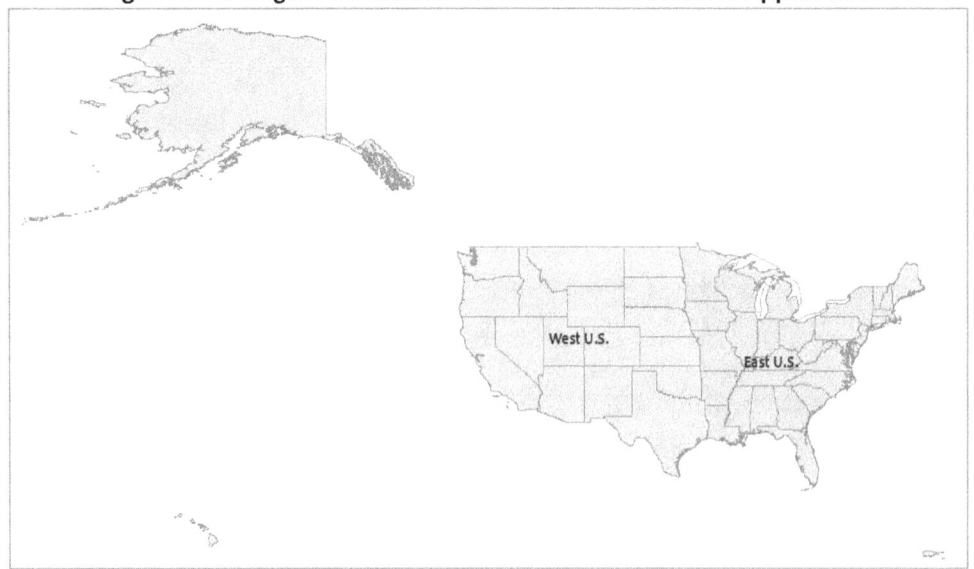

MTBS starts primarily from ICS-209 reports and solicits additional data from the states if inclusion in ICS-209 is unclear. Other sources for fire occurrence data include federal data, the National Fire Plan Operations and Reporting System (NFPORS), and InciWeb. These records are compiled into a standardized project database. MTBS identifies corresponding imagery using the Global Visualization Image Selection (GLOVIS) browser developed by the USGS EROS Center. ArcGIS shape files and scene-specific Advanced Very High Resolution Radiometer (AVHRR) greenness plots are incorporated into the viewer to aid scene selection and determination of peak periods of photosynthetic activity. Pre-fire and post-fire images are selected for each incident. Wildfires are analyzed on the scale of individual incidents, but the data can also be aggregated at other spatial scales for analytical purposes.

Figure 4. Land Area Burned by Wildfires by State, 1984–2012

Figure 4 is based on acreage data for large fires as compiled by the MTBS program through the analytical steps described above for Figure 4. These numbers are based on ICS-209 reports and additional state data compiled by MTBS.

6. Indicator Derivation

Figures 1 and 2. Wildfire Frequency and Acreage in the United States, 1983–2013

NIFC compiles local, state, and national reports to create the annual summary statistics published online. Data are aggregated to provide national, state and local statistics. EPA aggregated state-by-state totals in the annual Smokey Bear Reports to generate additional measures of annual wildfire frequency and extent.

Figure 3. Damage Caused by Wildfires in the United States, 1984–2012

Burn severity is a qualitative measure describing the degree to which a site has been altered by fire (NWCG, 2012). MTBS uses the Normalized Burn Ratio (NBR) to measure burn severity. NBR is a normalized index that uses satellite imagery from Landsat 5 and/or Landsat 7 TM/ETM bands 4 (near-infrared) and 7 (mid-infrared) to compare photosynthetically healthy and burned vegetation. Pre- and post-fire NBR are calculated to compare vegetation conditions before and after each wildfire.

The difference between pre- and post-fire NBRs is the Differenced Normalized Burn Ratio (dNBR). Calculated dNBR values are compared to established severity classes to give a qualitative assessment of the effects of fire damage. These classifications plus a full discussion of NBR and dNBR calculation methodology are described at: http://burnseverity.cr.usgs.gov/pdfs/LAv4_BR_CheatSheet.pdf.

Selected satellite images are also filtered through a complex sequence of data pre-processing, perimeter delineation, and other data quality assurance techniques. These procedures are documented in full on the MTBS website at: www.mtbs.gov/methods.html and in a 2005 report on western U.S. fires (MTBS, 2005).

The timing of the satellite imagery selected for analysis depends on the type of assessment that is conducted for a given fire. The optimal assessment type is selected based on the biophysical setting in which each fire occurs. MTBS conducts two main types of assessments:

- Initial Assessments compare imagery from shortly before and shortly after the fire, typically relying on the first available satellite data after the fire—on the scale of a few days. These assessments focus on the maximum post-fire data signal and are used primarily in ecosystems that exhibit rapid post-fire vegetation response (i.e., herbaceous and particular shrubland systems).

- Extended Assessments compare "peak green" conditions in the subsequent growing season with "peak green" conditions in the previous growing season, prior to the fire. These assessments are designed to capture delayed first-order effects (e.g., latent tree mortality) and dominant second-order effects that are ecologically significant (e.g., initial site response and early secondary effects).

MTBS occasionally conducts a Single Scene Assessment, which uses only a post-fire image (either "initial" or "extended"), when limited by factors such as data availability.

See: www.mtbs.gov/glossary.html for a glossary of MTBS assessment terms.

Figure 3 was created by filtering MTBS's database output to remove any fires not meeting MTBS's size criteria—although most such fires would not have been processed by MTBS anyway—and removing fires classified as "prescribed," "wildland fire use," or "unknown." The resulting analysis is therefore limited to fires classified as true "wildfires."

The total acreage shown in Figure 3 (the sum of the stacked burn severity sections) does not match the total acreage in Figure 2 because the burn severity analysis in Figure 3 is limited to fires above a specific size threshold (≥ 1,000 acres in the western United States and ≥ 500 acres in the eastern United States) and because the graph does not include acreage classified as "non-processing area mask," which denotes areas within the fire perimeter that could not be assessed for burn severity because the imagery was affected by clouds, cloud shadows, or data gaps. The Key Points text that describes the proportion of high severity acreage is based on high severity as a percentage of total assessed acreage (i.e., the total acreage after non-processing area has been excluded).

Figure 4. Land Area Burned by Wildfires by State, 1984–2012

Figure 4 presents two maps with state-level data: (a) normalized acreage burned per year and (b) the change in burned acreage over time.

To create map (a), EPA divided the annual acreage burned in each state by the state's total land area. After doing this for all years during the period of record (1984–2012), EPA calculated an average value and plotted it on the map.

To create map (b), EPA calculated each state's average annual acreage burned per square mile for the first half of the record (1984–1998) and the average for the second half (1999–2012). EPA found the difference between these values and expressed it as a percentage difference (e.g., average annual acreage during the second half of the record was 10 percent higher than average annual acreage burned during the first half). Changes have been characterized using this method rather than measuring a slope over time (e.g., a linear regression) because of the length and shape of the data set. Visual inspection of the NIFC line in Figure 2 (burned acreage across all states) suggests periods of relative stability punctuated by a noticeable jump in acreage during the late 1990s. This jump coincides with a period of transition in certain natural climate oscillations that tend to shift every few decades—notably, a shift in the Pacific Decadal Oscillation (PDO) around 1998 (Peterson and Schwing, 2003; Rodionov and Overland, 2005). This shift—combined with other ongoing changes in temperature, drought, and snowmelt—may have contributed to warmer, drier conditions that have fueled wildfires in parts of the western United States (Kitzberger et al., 2007; Westerling et al., 2006). With approximately 30 years of data punctuated by a phase transition, and with research strongly suggesting that the PDO and other decadal-scale oscillations contribute to cyclical patterns in wildfires in the western United States, EPA determined that linear regression is not an appropriate method of describing changes over time in this particular indicator. Instead, EPA chose to simply compare two sub-periods in a manner that considers all years of data and avoids inferring an annual rate of change. Without a nuanced statistical analysis to define a break point between two sub-periods, EPA chose to simply break the record into two halves of approximately equal length: 1984–1998 (15 years) and 1999–2012 (14 years). The fact that the break point currently lands at 1998 by this method is a coincidence. As more data are added in future years, the "halfway" break point will move accordingly.

EPA plans to investigate opportunities for a more robust interpretation of state-level trends over time in future editions of this indicator.

Comparison of Sources

Figure TD-2 compares total wildfire extent estimates from NIFC, Smokey Bear Reports, and MTBS. This graph shows that MTBS estimates follow the same pattern as the NIFC data set but are always somewhat lower than NIFC's totals because MTBS excludes small fires. The graph also shows how the most recent MTBS estimates compare with the previous MTBS data release. As expected, the data show evidence of revisions to historical data, but the changes are not extensive.

Figure TD-2. Comparison of Wildfire Extent from Three Data Sources, 1983–2013

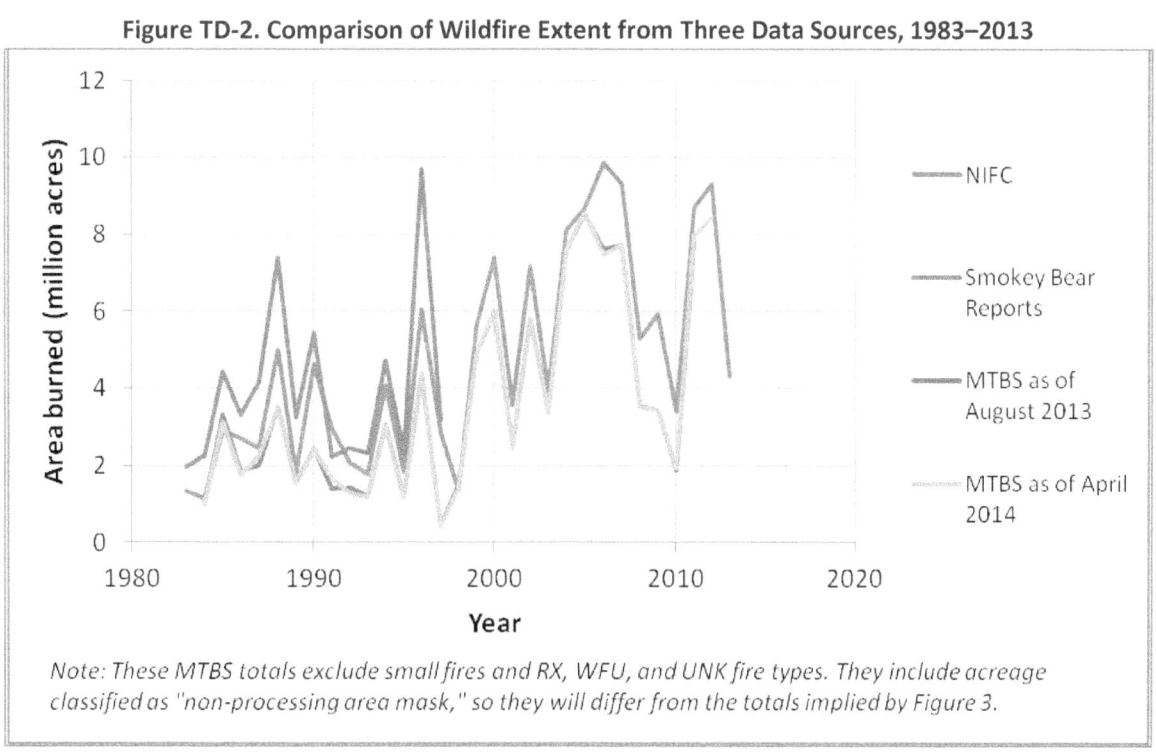

Note: These MTBS totals exclude small fires and RX, WFU, and UNK fire types. They include acreage classified as "non-processing area mask," so they will differ from the totals implied by Figure 3.

NIFC's website provides data from 1960 through 2013, and Smokey Bear Reports are available from 1917 to 1997. However, the data available prior to the early 1980s provide incomplete geographic coverage, as fire statistics at the time were not compiled from the full extent of "burnable" lands. Thus, Figures 1 and 2 of this indicator begin in 1983, which was the first year of nationwide reporting via ICS-209 reports. Figures 3 and 4 begin in 1984, which was the first year for which the MTBS project conducted its detailed analysis. MTBS depends on aerial imagery and the level of detail captured consistently in ICS reports. Thus, while a longer period of record would be desirable when analyzing long-term changes in a climatological context, EPA could not extend this indicator with pre-1983 data without introducing inconsistencies and gaps that would preclude meaningful comparisons over time and space.

For more discussion regarding the availability, coverage, and reliability of historical wildfire statistics, see the authoritative discussion in Short (2013). An accompanying publicly available seminar (http://videos.firelab.org/ffs/2013-14Seminar/032014Seminar/032014Seminar.html) explains additional nuances and advises users on how they should and should not use the data. Based on these sources, Table TD-1 summarizes the available data sets, their coverage, and their underlying sources. NIFC's pre-1983 estimates actually derive from the Smokey Bear Reports (Karen Short, USDA Forest Service: personal communication and publicly available seminar cited above); therefore, in reality, the Smokey Bear Reports are the only underlying nationwide source of pre-1983 wildfire statistics.

Table TD-1. Comparison of Historical Wildfire Data Sources

Data set	Variables	Temporal range	Resolution	Geographic coverage	Underlying sources
NIFC *(Figures 1 and 2)*	Acreage and incidence (number of fires)	1983–2013	Annual	National	ICS incident reports
NIFC pre-1983	Acreage and incidence	1960–1982	Annual	National with gaps	Smokey Bear Reports, which are based on estimates submitted by various agencies
Smokey Bear Reports *(recent data in Figures 1 and 2)*	Acreage and incidence	1917–1997	Annual	National with gaps	Estimates submitted by various agencies
MTBS *(Figures 3 and 4)*	Burn severity; acreage by state	1984–2012	Annual	National	ICS incident reports

A fundamental shift in wildfire reporting took place in the early 1980s with the onset of the ICS reporting system. Prior to this time, reports were submitted to the USDA Forest Service by selected state and federal agencies, covering land within each agency's jurisdiction. Many of these reports were limited to fires on land with "protected" status (i.e., land designated for cooperative fire control). Fires occurring on "unprotected" land would not necessarily be fought, and they would not be counted in the statistics either. Figure TD-3 below, based on data obtained from the USDA Forest Service (Karen Short), demonstrates how the reporting area was well below the total nationwide "burnable" acreage until the 1980s. Increases in the reporting area occurred when additional agencies joined the reporting program. For example, the Bureau of Land Management began reporting around 1964, which accounts for the noticeable jump in the blue line in Figure TD-3.

Figure TD-3. "Smokey Bear Reports" Reporting Area Over Time

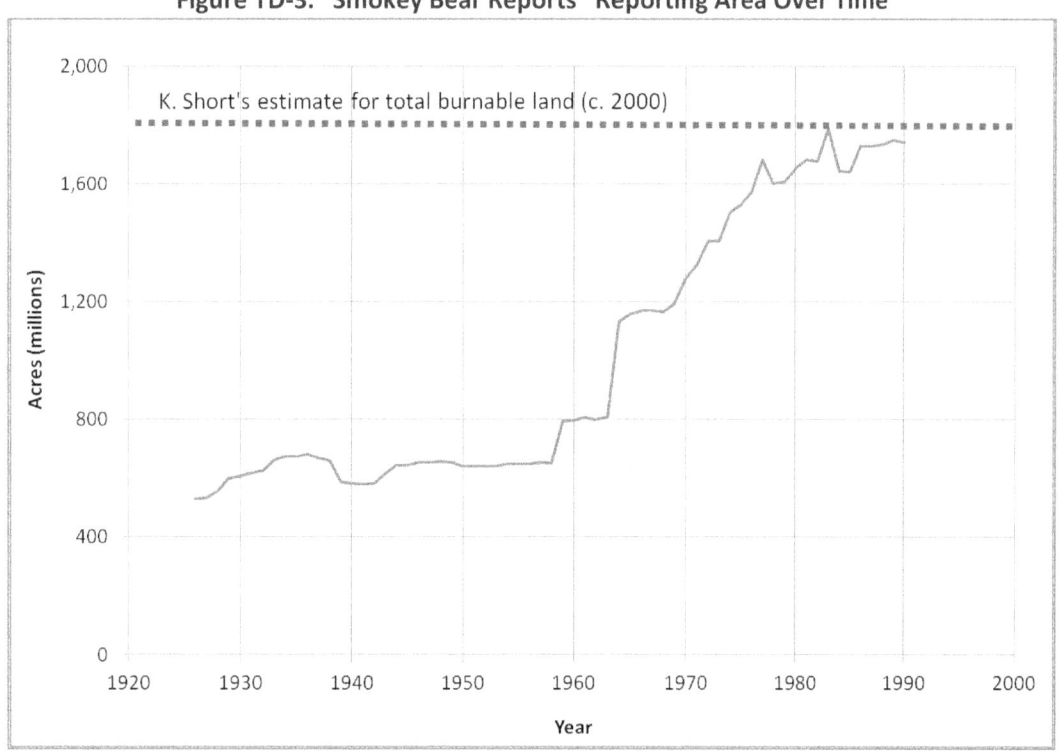

The Smokey Bear Reports achieved essentially complete participation and coverage by the early 1980s. They continued to be compiled even with the advent of the ICS system around 1983, until being phased out in the late 1990s. Thus, the Smokey Bear Reports and the ICS reports provide complementary coverage for much of the 1980s and 1990s. During the first few years of ICS data collection—particularly 1983 and 1984—the Smokey Bear Reports showed much higher fire occurrence than NIFC's ICS-derived statistics. The USDA Forest Service attributes this difference to the ramp-up and gradual adoption of ICS nationwide. Other studies such as Dennison et al. (2014) also describe the advantages of using recent, robust data sets instead of longer, less complete fire databases previously used for trend analysis.

7. Quality Assurance and Quality Control

The ICS-209 form provides clear guidelines for wildfire reporting for state, local, and federal agencies. These guidelines are accessible on the NIFC website at: http://gacc.nifc.gov/nrcc/dc/idgvc/dispatchforms/ics209.tips.pdf. The information in the ICS-209 forms is compiled by the SIT program to provide daily situation reports that summarize wildfire conditions at state and national levels. This compiled information forms the basis for NIFC summary statistics. The NIFC does not provide details on how it counts, measures acreage, or filters out the double reporting of fires, fires that split, fires that merge, incomplete forms, or other potential data irregularities. However, the frequency of these confounding factors is likely limited and may not seriously compromise the quality of the data presented in this indicator.

MTBS standardizes and corrects raw fire incidence data. The project avoids editing source data with the exception of correcting a record's geospatial coordinates if (1) the coordinates are clearly incorrect, and (2) a correction can be made with confidence. All selected scenes from the database are ordered and processed following existing USGS-EROS protocols. Data obtained from MTBS were also cross-checked prior to conducting analyses for this indicator.

Analysis

8. Comparability Over Time and Space

NIFC methods and statistics have not changed since 1983, and they can be compared on an annual basis at national scales. The sole exception is the NIFC fire count and burned acreage data points for the year 2004, which are missing totals from state lands in North Carolina. Thus, these two points slightly compromise the comparability over both time and space. Smokey Bear Reports also used consistent methods from 1983 to 1997, and they covered the full extent of "burnable" U.S. lands throughout this period. MTBS has used consistent methods to classify burn severity from 1984 through 2012, allowing for annual comparisons through time and allowing for spatial comparisons among states. MTBS is based on a type of satellite imagery that has been collected consistently with sufficient resolution to support this analysis throughout the period of record.

Figure 4 was derived from an MTBS data set that uses different size thresholds for different states. This data set includes fires ≥ 1,000 acres in the western United States and ≥ 500 acres in the eastern United States. Thus, the map might undercount small fires in the West, compared with the East. These thresholds have held consistent for each state over time, however, which lends validity to the analysis of state-level trends over time.

9. Data Limitations

Factors that may impact the confidence, application, or conclusions drawn from this indicator are as follows:

1. Wildfire activity can be influenced by a variety of other factors besides climate. Examples include changes in human activities and land management strategies over time, particularly changes in fire suppression and fire management practices, which (among other things) can potentially contribute to more damaging fires in the future if they result in a buildup of fuel in the understory. Grazing activities can also influence the amount and type of vegetation in the landscape, and changes in land cover and land use—for example, forest to non-forest conversion—can affect the extent and type of "burnable" land. Thus, further analysis is needed before an apparent change in wildfire activity can necessarily be attributed to climate change.

2. The dominant drivers of wildfire activity can vary by region. Contributing factors may include (but are not limited to) temperatures in specific seasons (particularly spring), drought, and precipitation that contributes to vegetation growth. As described in Section 6, wildfire trends in

some regions have been linked with certain phases of multi-year and multi-decadal climate oscillations (Kitzberger et al., 2007; Westerling et al., 2006). Climate patterns that lead to more wildfire activity in some parts of the United States may lead to a simultaneous decrease in activity in other regions (e.g., the Northwest versus the Southwest). Reconstructions based on tree rings can provide hundreds of years of context for understanding such patterns and how they vary regionally (e.g., Swetnam and Betancourt, 1998).

3. While this indicator is officially limited to "wildland" fires, it includes fires that encroach on—or perhaps started in—developed areas at the wildland-urban interface (WUI). Encroachment of the WUI over time into previously wild lands could influence trends in wildfire frequency and extent (Radeloff et al., 2005).

4. NIFC data, which are derived from government entities of varying scope or jurisdiction, can be limited by inconsistencies across how data are reported through ICS-209 forms. With aggregation from so many potential sources, wildfire incidence data, particularly historical data, may be redundant or erroneous. Data aggregation among sources may result in variability in reporting accuracy and protocol.

5. The MTBS program depends on certain conditions to make accurate measurements of burn severity:

   • Accurate fire location coordinates that match burn scars visible via satellite.
   • Accurate fire size information that ensures that fires meeting the MTBS size criteria are properly included.
   • Accurate date of ignition and out date that guide the appropriate selection of imagery, particularly for baseline assessments.
   • Pre-fire and post-fire images that are cloud-free to avoid visual obscuration of the fire area.

6. Some fires of very low severity may not be visible in the satellite imagery and therefore impossible to delineate or characterize. Cloud cover, cloud shadow, or data gaps can also preclude damage assessment. To account for all of these limitations, the MTBS project includes a burn severity classification of "non-processing area mask." This classification accounts for approximately 5.4 percent of the total wildfire acreage from 1984 through 2012.

10. Sources of Uncertainty

Uncertainties in these data sets have not been quantified. The most likely sources of uncertainty relate to initial data collection methods. Federal land management agencies have varying standards for content, geospatial accuracy, and nomenclature. Duplicate records occur due to reporting of a given incident by multiple agencies, such as redundant reports from local, state, or federal entities. In any given year, as much as three-quarters of all fire incidents are reported by non-federal state and local agencies (NICC, 2012). Cases of gross geospatial inaccuracies may also occur. Similar inconsistencies occur within state databases. However, the MTBS project addresses issues such as duplicates and nomenclature during pre-processing.

11. Sources of Variability

Forest conditions, and therefore wildfire incidents, are highly affected by climate conditions. In addition to year-to-year variations, evidence suggests that wildfire patterns in the western United States are influenced by multi-year and multi-decadal climate oscillations such as the PDO (http://jisao.washington.edu/pdo) and the Atlantic Multidecadal Oscillation (www.aoml.noaa.gov/phod/amo_faq.php). For example, see Kitzberger et al. (2007) and Westerling et al. (2006) for discussion of warmer, drier conditions that have contributed to increases in wildfire activity in certain regions.

Changes in the frequency of wildfire triggers (e.g., lightning, negligent or deliberate human activity) could also affect wildfire frequency. Burn severity is affected by local vegetation regimes and fuel loads. Finally, improvements or strategic changes in firefighting and fire management may affect wildfire prevalence and resulting damages. Forest management practices have changed over time from complete fire suppression to controlled burns. These varied approaches and the extent to which they are applied on state or regional levels can influence the wildfire data presented in this indicator.

12. Statistical/Trend Analysis

As described in Section 6, the nature of this topic and the length and shape of the time series suggest that linear regression is not a suitable tool for characterizing long-term trends and their significance. Thus, the figures and Key Points do not report regression results. Ordinary least-squares linear regressions from the NIFC data set have been calculated here, just for reference. Regression slopes and p-values are indicated in Table TD-2 below.

Table TD-2. Wildfire Regression Statistics

Indicator component	Regression slope	P-value
NIFC fire frequency (Figure 1)	+552 fires/year	0.14
NIFC burn acreage (Figure 2)	+203,322 acres/year	<0.001

References

Dennison, P.E., S.C. Brewer, J.D. Arnold, and M.A. Moritz. 2014. Large wildfire trends in the western United States, 1984–2011. Geophys. Res. Lett. (online pre-print).

Fire and Aviation Management and Predictive Services. 2009. Interagency wildland fire incident information reporting application (SIT-209). www.predictiveservices.nifc.gov/intelligence/SIT-209_Bus_Req_Final_v1a.pdf.

Kitzberger, T., P.M. Brown, E.K. Heyerdahl, T.W. Swetnam, and T.T. Veblen. 2007. Contingent Pacific–Atlantic Ocean influence on multicentury wildfire synchrony over western North America. P. Natl. Acad. Sci. USA 104(2):543–548.

MTBS (Monitoring Trends in Burn Severity). 2005. Report on the Pacific Northwest and Pacific Southwest fires. www.mtbs.gov/files/MTBS_pnw-psw_final.pdf.

NICC (National Interagency Coordination Center). 2012. Wildland fire summary and statistics annual report. www.predictiveservices.nifc.gov/intelligence/2012_statssumm/annual_report_2012.pdf.

NWCG (National Wildfire Coordinating Group). 2012. Glossary of wildland fire terminology. www.nwcg.gov/pms/pubs/glossary/index.htm.

Peterson, W.T., and F.B. Schwing. 2003. A new climate regime in northeast Pacific ecosystems. Geophys. Res. Lett. 30(17).

Radeloff, V.C., R.B. Hammer, S.I. Stewart, J.S. Fried, S.S. Holcomb, and J.F. McKeefry. 2005. The wildland-urban interface in the United States. Ecol. Appl. 15:799–805.

Rodionov, S., and J.E. Overland. 2005. Application of a sequential regime shift detection method to the Bering Sea ecosystem. ICES J. Mar. Sci. 62:328–332.

Short, K.C. 2013. A spatial database of wildfires in the United States, 1992–2011. Earth Syst. Sci. Data Discuss. 6:297–366. www.fs.fed.us/rm/pubs_other/rmrs_2013_short_k001.pdf.

Swetnam, T.W., and J.L. Betancourt. 1998. Mesoscale disturbance and ecological response to decadal climatic variability in the American Southwest. J. Climate 11:3128–3147.

Westerling, A.L., H.G. Hidalgo, D.R. Cayan, and T.W. Swetnam. 2006. Warming and earlier spring increase western U.S. forest wildfire activity. Science 313(5789):940–943.

Streamflow

Identification

1. Indicator Description

This indicator describes trends in the magnitude and timing of streamflow in streams across the United States. Streamflow is a useful indicator of climate change for several reasons. Changes in the amount of snowpack and earlier spring melting can alter the size and timing of peak streamflows. More precipitation is expected to cause higher average streamflow in some places, while heavier storms could lead to larger peak flows. More frequent or severe droughts will reduce streamflow in certain areas.

Components of this indicator include trends in four annual flow statistics:

- Magnitude of annual seven-day low streamflow from 1940 through 2012 (Figure 1).
- Magnitude of annual three-day high streamflow from 1940 through 2012 (Figure 2).
- Magnitude of annual mean streamflow from 1940 through 2012 (Figure 3).
- Timing of winter-spring center of volume date from 1940 through 2012 (Figure 4).

2. Revision History

December 2011:   Indicator developed.
April 2012:   Updated with a new analysis.
December 2013:   Original figures updated with data through 2012; new Figure 3 (annual mean
                    streamflow) added; original Figure 3 (winter-spring center of volume)
                    renumbered as Figure 4.

Data Sources

3. Data Sources

This indicator was developed by Mike McHale, Robert Dudley, and Glenn Hodgkins at the U.S. Geological Survey (USGS). The indicator is based on streamflow data from a set of reference stream gauges specified in the Geospatial Attributes of Gages for Evaluating Streamflow (GAGES-II) database, which was developed by USGS and is described in Lins (2012). Daily mean streamflow data are housed in the USGS National Water Information System (NWIS).

4. Data Availability

EPA obtained the data for this indicator from Mike McHale, Robert Dudley, and Glenn Hodgkins at USGS. Similar streamflow analyses had been previously published in the peer-reviewed literature (Burns et al., 2007; Hodgkins and Dudley, 2006). The USGS team provided a reprocessed data set to include streamflow trends through 2012.

Streamflow data from individual stations are publicly available online through the surface water section of NWIS at: http://waterdata.usgs.gov/nwis/sw. Reference status and watershed, site characteristics, and other metadata for each stream gauge in the GAGES-II database are available online at: http://water.usgs.gov/GIS/metadata/usgswrd/XML/gagesII_Sept2011.xml.

Methodology

5. Data Collection

Streamflow is determined from data collected at stream gauging stations by devices that record the elevation (or stage) of a river or stream at regular intervals each day. USGS maintains a national network of stream gauging stations, including more than 7,000 stations currently in operation throughout the United States (http://water.usgs.gov/wid/html/SG.html). USGS has been collecting stream gauge data since the late 1800s at some locations. Gauges generally are sited to record flows for specific management or legal issues, typically in cooperation with municipal, state, and federal agencies. Stream surface elevation is recorded at regular intervals that vary from station to station—typically every 15 minutes to one hour.

Streamflow (or discharge) is measured at regular intervals by USGS personnel (typically every four to eight weeks). The relation between stream stage and discharge is determined and a stage-discharge relation (rating) is developed to calculate streamflow for each recorded stream stage (Rantz et al., 1982). These data are used to calculate the daily mean discharge for each day at each site. All measurements are made according to standard USGS procedures (Rantz et al., 1982; Sauer and Turnipseed, 2010; Turnipseed and Sauer, 2010).

This indicator uses data from a subset of USGS stream gauges that have been designated as Hydro-Climatic Data Network (HCDN)-2009 "reference gauges" (Lins, 2012). These reference gauges have been carefully selected to reflect minimal interference from human activities such as dam construction, reservoir management, wastewater treatment discharge, water withdrawal, and changes in land cover and land use that might influence runoff. The subset of reference gauges was further winnowed on the basis of length of period of record (73 years) and completeness of record (greater than or equal to 80 percent for every decade). Figures 1, 2, and 3 are based on data from 193 stream gauges. Figure 4 relies on 56 stream gauges because it is limited to watersheds that receive 30 percent or more of their total annual precipitation in the form of snow. This additional criterion was applied because the metric in Figure 4 is used primarily to examine the timing of winter-spring runoff, which is substantially affected by snowmelt-related runoff in areas with a large annual snowpack. All of the selected stations and their corresponding basins are relatively independent—that is, the analysis does not include gauges with substantially overlapping watershed areas.

All watershed characteristics, including basin area, station latitude and longitude, and percentage of precipitation as snow were taken from the GAGES-II database. GAGES-II basin area was determined through EPA's National Hydrography Dataset Plus and supplemented by the USGS National Water-Quality Assessment Program and the USGS Elevation Derivatives for National Applications.

6. Indicator Derivation

Figures 1, 2, and 3. Seven-Day Low (Figure 1), Three-Day High (Figure 2), and Annual Average (Figure 3) Streamflow in the United States, 1940–2012

Figure 1 shows trends in low-flow conditions using seven-day low streamflow, which is the lowest average of seven consecutive days of streamflow in a calendar year. Hydrologists commonly use this measure because it reflects sustained dry or frozen conditions that result in the lowest flows of the year. Seven-day low flow can equal zero if a stream has dried up completely.

Figure 2 shows trends in wet conditions using three-day high streamflow, which is the highest average of three consecutive days of streamflow in a calendar year. Hydrologists use this measure because a three-day averaging period has been shown to effectively characterize runoff associated with large storms and peak snowmelt over a diverse range of watershed areas.

Figure 3 shows trends in average conditions using annual mean streamflow, which is the average of all daily mean streamflow values for a given calendar year.

Rates of change from 1940 to 2012 at each station on the maps were computed using the Sen slope, which is the median of all possible pair-wise slopes in a temporal data set (Helsel and Hirsch, 2002). The Sen slope was then multiplied by the length of the period of record (72 years: the last year minus the starting year) to estimate total change over time. Trends are reported as percentage increases or decreases, relative to the beginning Sen-slope value.

Figure 4. Timing of Winter-Spring Runoff in the United States, 1940–2012

Figure 4 shows trends in the timing of streamflow in the winter and spring, which is influenced by the timing of snowmelt runoff in areas with substantial annual snowpack. The timing of streamflow also can be influenced by the ratio of winter rain to snow and by changes in the seasonal distribution of precipitation. The measurement in Figure 4 uses the winter-spring center of volume (WSCV) date, which is defined for this indicator as the date when half of the total streamflow that occurred between January 1 and June 30 has passed by the gauging station. Trends in this date are computed in the same manner as the other three components of this indicator, and the results are reported in terms of the number of days earlier or later that WSCV is occurring. For more information about WSCV methods, see Hodgkins and Dudley (2006) and Burns et al. (2007).

7. Quality Assurance and Quality Control

Quality assurance and quality control (QA/QC) procedures are documented for measuring stream stage (Sauer and Turnipseed, 2010), measuring stream discharge (Turnipseed and Sauer, 2010), and computing stream discharge (Sauer, 2002; Rantz et al., 1982). Stream discharge is typically measured and equipment is inspected at each gauging station every four to eight weeks. The relation between stream stage and stream discharge is evaluated following each discharge measurement at each site, and shifts to the relation are made if necessary.

The GAGES-II database incorporated a QC procedure for delineating the watershed boundaries acquired from the National Hydrography Dataset Plus. The data set was cross-checked against information from USGS's National Water-Quality Assessment Program. Basin boundaries that were inconsistent across sources were visually compared and manually delineated based on geographical information provided in USGS's Elevation Derivatives for National Applications. Other screening and data quality issues are addressed in the GAGES-II metadata available at: http://water.usgs.gov/GIS/metadata/usgswrd/XML/gagesII_Sept2011.xml.

Analysis

8. Comparability Over Time and Space

All USGS streamflow data have been collected and extensively quality-assured by USGS since the start of data collection. Consistent and well-documented procedures have been used for the entire periods of recorded streamflows at all gauges (Corbett et al., 1943; Rantz et al., 1982; Sauer, 2002).

Trends in streamflow over time can be heavily influenced by human activities upstream, such as the construction and operation of dams, flow diversions and abstractions, and land use change. To remove these artificial influences to the extent possible, this indicator relies on a set of reference gauges that were chosen because they represent least-disturbed (though not necessarily completely undisturbed) watersheds. The criteria for selecting reference gauges vary from region to region based on land use characteristics. This inconsistency means that a modestly impacted gauge in one part of the country (e.g., an area with agricultural land use) might not have met the data quality standards for another less impacted region. The reference gauge screening process is described in Lins (2012) and is available in the GAGES-II metadata at: http://water.usgs.gov/GIS/metadata/usgswrd/XML/gagesII_Sept2011.xml.

Analytical methods have been applied consistently over time and space.

9. Data Limitations

Factors that may impact the confidence, application, or conclusions drawn from this indicator are as follows:

1. This analysis is restricted to locations where streamflow is not highly disturbed by human influences, including reservoir regulation, diversions, and land cover change. However, changes in land cover and land use over time could still influence trends in the magnitude and timing of streamflow at some sites.

2. Reference gauges used for this indicator are not evenly distributed throughout the United States, nor are they evenly distributed with respect to topography, geology, elevation, or land cover.

3. Some streams in northern or mountainous areas have their lowest flows in the winter due to water being held in snow or ice for extended periods. As a result, their low flow trends could be influenced by climate factors other than reduced precipitation or otherwise dry conditions.

10. Sources of Uncertainty

Uncertainty estimates are not available for this indicator as a whole. As for the underlying data, the precision of individual stream gauges varies from site to site. Accuracy depends primarily on the stability of the stage-discharge relationship, the frequency and reliability of stage and discharge measurements, and the presence of special conditions such as ice (Novak, 1985). Accuracy classifications for all USGS gauges for each year of record are available in USGS annual state water data reports. USGS has published a general online reference devoted to the calculation of error in individual stream discharge measurements (Sauer and Meyer, 1992).

11. Sources of Variability

Streamflow can be highly variable over time, depending on the size of the watershed and the factors that influence flow at a gauge. USGS addresses this variability by recording stream stage many times a day (typically 15-minute to one-hour intervals) and then computing a daily average streamflow. Streamflow also varies from year to year as a result of variation in precipitation and air temperature. Trend magnitudes computed from Sen slopes provide a robust estimate of linear changes over a period of record, and thus this indicator does not measure decadal cycles or interannual variability in the metric over the time period examined.

While gauges are chosen to represent drainage basins relatively unimpacted by human disturbance, some sites may be more affected by direct human influences (such as land-cover and land-use change) than others. Other sources of variability include localized factors such as topography, geology, elevation, and natural land cover. Changes in land cover and land use over time can contribute to streamflow trends, though careful selection of reference gauges strives to minimize these impacts.

Although WSCV is driven by the timing of the bulk of snow melt in areas with substantial annual snowpack, other factors also will influence WSCV. For instance, a heavy rain event in the winter could result in large volumes of water that shift the timing of the center of volume earlier. Changes over time in the distribution of rainfall during the January–June period could also affect the WSCV date.

12. Statistical/Trend Analysis

The maps in Figures 1, 2, 3, and 4 all show trends through time that have been computed for each gauging station using a Sen slope analysis. Because of uncertainties and complexities in the interpretation of statistical significance, particularly related to the issue of long-term persistence (Cohn and Lins, 2005; Koutsoyiannis and Montanari, 2007), significance of trends is not reported.

References

Burns, D.A., J. Klaus, and M.R. McHale. 2007. Recent climate trends and implications for water resources in the Catskill Mountain region, New York, USA. J. Hydrol. 336(1–2):155–170.

Cohn, T.A., and H.F. Lins. 2005. Nature's style: Naturally trendy. Geophys. Res. Lett. 32:L23402.

Corbett, D.M., et al. 1943. Stream-gaging procedure: A manual describing methods and practices of the Geological Survey. U.S. Geological Survey Water-Supply Paper 888. http://pubs.er.usgs.gov/publication/wsp888.

Helsel, D.R., and R.M. Hirsch. 2002. Statistical methods in water resources. Techniques of water resources investigations, book 4. Chap. A3. U.S. Geological Survey. http://pubs.usgs.gov/twri/twri4a3.

Hodgkins, G.A., and R.W. Dudley. 2006. Changes in the timing of winter-spring streamflows in eastern North America, 1913–2002. Geophys. Res. Lett. 33:L06402. http://water.usgs.gov/climate_water/hodgkins_dudley_2006b.pdf.

Koutsoyiannis, D., and A. Montanari. 2007. Statistical analysis of hydroclimatic time series: Uncertainty and insights. Water Resour. Res. 43(5):W05429.

Lins, H.F. 2012. USGS Hydro-Climatic Data Network 2009 (HCDN-2009). U.S. Geological Survey Fact Sheet 2012-3047. http://pubs.usgs.gov/fs/2012/3047.

Novak, C.E. 1985. WRD data reports preparation guide. U.S. Geological Survey Open-File Report 85-480. http://pubs.er.usgs.gov/publication/ofr85480.

Rantz, S.E., et al. 1982. Measurement and computation of streamflow. Volume 1: Measurement of stage and discharge. Volume 2: Computation of discharge. U.S. Geological Survey Water Supply Paper 2175. http://pubs.usgs.gov/wsp/wsp2175.

Sauer, V.B. 2002. Standards for the analysis and processing of surface-water data and information using electronic methods. U.S. Geological Survey Water-Resources Investigations Report 01-4044. http://pubs.er.usgs.gov/publication/wri20014044.

Sauer, V.B., and R.W. Meyer. 1992. Determination of error in individual discharge measurements. U.S. Geological Survey Open-File Report 92-144. http://pubs.usgs.gov/of/1992/ofr92-144.

Sauer, V.B., and D.P. Turnipseed. 2010. Stage measurement at gaging stations. U.S. Geological Survey Techniques and Methods book 3. Chap. A7. U.S. Geological Survey. http://pubs.usgs.gov/tm/tm3-a7.

Turnipseed, D.P., and V.P. Sauer. 2010. Discharge measurements at gaging stations. U.S. Geological Survey Techniques and Methods book 3. Chap. A8. U.S. Geological Survey. http://pubs.usgs.gov/tm/tm3-a8.

Great Lakes Water Levels and Temperatures

Identification

1. Indicator Description

This indicator describes how water levels and surface water temperatures in the Great Lakes (Lake Superior, Lake Michigan, Lake Huron, Lake Erie, and Lake Ontario) have changed over the last 150 years (water levels) and the last two decades (temperatures). Water levels and surface water temperatures are useful indicators of climate change because they can be affected by air temperatures, precipitation patterns, evaporation rates, and duration of ice cover. In recent years, warmer surface water temperatures in the Great Lakes have contributed to lower water levels by increasing rates of evaporation and causing lake ice to form later than usual, which extends the season for evaporation (Gronewold et al., 2013).

Components of this indicator include:

- Average annual water levels in the Great Lakes since 1860 (Figure 1).
- Average annual surface water temperatures of the Great Lakes since 1995 (Figure 2).
- Comparison of daily surface water temperatures throughout the year, 1995–2004 versus 2005–2013 (Figure 2).

2. Revision History

December 2013:   Indicator proposed.
April 2014:   Updated with data through 2013.

Data Sources

3. Data Sources

Water level data were collected by water level gauges and were provided by the National Oceanic and Atmospheric Administration's (NOAA's) National Ocean Service (NOS), Center for Operational Oceanographic Products and Services (CO-OPS) and the Canadian Hydrographic Service (CHS). Water level data are available for the period 1860 to 2013.

The temperature component of this indicator is based on surface water temperature data from satellite imagery analyzed by NOAA's Great Lakes Environmental Research Laboratory's Great Lakes Surface Environmental Analysis (GLSEA). Complete years of satellite data are available from 1992 to 2013.

4. Data Availability

All of the Great Lakes water level and surface temperature observations used for this indicator are publicly available from the following NOAA websites:

- Water level data from the Great Lakes Water Level Dashboard Data Download Portal: www.glerl.noaa.gov/data/now/wlevels/dbd/GLWLDDataDownloads2.html.

- Water level data documentation: www.glerl.noaa.gov/data/now/wlevels/dbd/levels.html.

- Satellite-based temperature data from GLSEA: http://coastwatch.glerl.noaa.gov/ftp/glsea/avgtemps.

Methodology

5. Data Collection

Water Levels

NOAA's NOS/CO-OPS and CHS use a set of gauges along the shoreline to measure water levels in each of the five Great Lakes. All five lakes have had one or more gauges in operation since 1860. In 1992, the Coordinating Committee for Great Lakes Basic Hydraulic and Hydrologic Data approved a standard set of gauges suitable for both U.S. and Canadian shores, covering the period from 1918 to present. These gauges were chosen to provide the most accurate measure of each lake's water level when averaged together. The standard set comprises 22 gauges in the five Great Lakes and two in Lake St. Clair (the smaller lake between Lake Huron and Lake Erie). Only the five Great Lakes are included in this indicator. Lakes Michigan and Huron are combined for this analysis because they are hydrologically connected, and thus they are expected to exhibit the same water levels.

The locations of the water level gauges used for this indicator are shown in Table TD-1.

Table TD-1. Water Level Gauge Locations

Lake Superior	Lakes Michigan-Huron	Lake Erie	Lake Ontario
Duluth, MN	Ludington, MI	Toledo, OH	Rochester, NY
Marquette C.G., MI	Mackinaw City, MI	Cleveland, OH	Oswego, NY
Pt Iroquois, MI	Harbor Beach, MI	Fairport, OH	Port Weller, ON
Michipicoten, ON	Milwaukee, WI	Port Stanley, ON	Toronto, ON
Thunder Bay, ON	Thessalon, ON	Port Colborne, ON	Cobourg, ON
	Tobermory, ON		Kingston, ON

An interactive map of all CO-OPS stations is available online at: http://tidesandcurrents.noaa.gov/gmap3. For more information about data collection methods and the low water datum that is used as a reference plane for each lake, see: www.glerl.noaa.gov/data/now/wlevels/dbd/levels.html.

Surface Water Temperatures

The GLSEA is operated by NOAA's Great Lakes Environmental Research Laboratory through the NOAA CoastWatch program. For general information about this program, see: http://coastwatch.glerl.noaa.gov/glsea/doc. GLSEA uses data from the Polar-Orbiting Operational

Environmental Satellites system. Specifically, GLSEA uses data from the Advanced Very High Resolution Radiometer instrument, which can measure surface temperatures. Visit: www.ospo.noaa.gov/Operations/POES/index.html for more information about the satellite missions and: http://noaasis.noaa.gov/NOAASIS/ml/avhrr.html for detailed documentation of instrumentation. GLSEA satellite-based data for the Great Lakes are available from 1992 through the present. However, data for winter months in 1992 through 1994 are absent. Complete years of satellite-based data are available starting in 1995.

6. Indicator Derivation

Water Levels

NOAA provides annual average water level observations in meters, along with the highest and lowest monthly average water levels for each year. As discussed in Section 8, data provided for the period before 1918 represent observations from a single gauge per lake. NOAA corrected pre-1918 data for Lakes Superior and Erie to represent outlet water levels. NOAA averaged observations from multiple gauges per lake in the data from 1918 to present, using the standard set of gauges described in Section 5.

In Figure 1, water level data are presented as trends in anomalies to depict change over time. An anomaly represents the difference between an observed value and the corresponding value from a baseline period. This indicator uses a baseline period of 1981 to 2010, which is consistent with the 30-year climate normal used in many other analyses by NOAA and others in the scientific community. The choice of baseline period will not affect the shape or the statistical significance of the overall trend in anomalies. In this case, a different baseline would only move the time series up or down on the graph in relation to the point defined as zero. Water level anomalies were converted from meters to feet. The lines in Figure 1 show the annual average for each lake, while the shaded bands show the range of monthly values within each year.

Surface Water Temperatures

Surface water temperature observations are provided daily by satellite imagery. The left side of Figure 2 shows annual averages, which were calculated using arithmetic means of the daily satellite data. The right side of Figure 2 shows the pattern of daily average satellite-based temperatures over the course of a year. To examine recent changes, Figure 2 divides the record approximately in half and compares daily conditions averaged over 2005–2013 with daily conditions for the period 1995–2004. All temperatures were converted from Celsius to Fahrenheit to make them consistent with all of EPA's other temperature-related indicators.

General Notes

EPA did not attempt to interpolate missing data points. This indicator also does not attempt to portray data beyond the time periods of observation or beyond the five lakes that were selected for the analysis.

7. Quality Assurance and Quality Control

Water Levels

Lake-wide average water levels are calculated using a standard set of gauges established by the Coordinating Committee for Great Lakes Basic Hydraulic and Hydrologic Data in 1992. Data used in this indicator are finalized data, subject to internal quality assurance/quality control (QA/QC) standards within NOAA/NOS and CHS. Each gauge location operated by NOAA houses two water level sensors: a primary sensor and a redundant sensor. If data provided by the primary and redundant sensors differ by more than 0.003 meters, the sensors are manually checked for accuracy. In addition, a three standard deviation outlier rejection test is applied to each measurement, and rejected values are not included in calculated values.

Surface Water Temperatures

NOAA's National Data Buoy Center, which collects the buoy surface temperature observations, follows a comprehensive QA/QC protocol, which can be found in the Handbook of Automated Data Quality Control Checks and Procedures:
www.ndbc.noaa.gov/NDBCHandbookofAutomatedDataQualityControl2009.pdf.

Satellite observations of surface temperature are subject to several QA/QC measures prior to publication. All satellite data are validated by NOAA personnel. Following this step, an automated algorithm flags and excludes temperatures not in the normal range of expected temperatures, correcting for processing errors in the original satellite data. Finally, multiple cloud masks are applied to both day and night satellite imagery so that the final product includes only completely cloud-free data. An additional algorithm is used to correct for missing pixels. Two iterations of this algorithm are described in a presentation entitled "Overview of GLSEA vs. GLSEA2 [ppt]" at:
http://coastwatch.glerl.noaa.gov/glsea/doc.

Analysis

8. Comparability Over Time and Space

Water level observations prior to 1918 have been processed differently from those collected from 1918 to present. Prior to 1918, there were fewer water level gauges in the Great Lakes. As such, values from 1860 to 1917 represent one gauge per lake, which may not represent actual lake-wide average water levels. Corrections to data have been made to allow comparability over time. These corrections include adjustments due to the slow but continuing rise of the Earth's crust (including the land surface and lake bottoms) as a result of the retreat of the ice sheets after the last glacial maximum (commonly referred to as the last ice age), as well as adjustments to account for the relocation of gauges. For more discussion about these corrections, see: www.glerl.noaa.gov/data/now/wlevels/dbd/levels.html.

Satellite temperature observations have been made systematically since 1992, allowing for comparability over time. This indicator starts in 1995, which was the first year with complete coverage of all months for all lakes.

9. Data Limitations

Factors that may impact the confidence, application, or conclusions drawn from this indicator are as follows:

1. Besides climate change, natural year-to-year variability and other factors such as human use and contamination can influence water temperatures.

2. Satellite data are only available starting in 1992, and the years 1992–1994 were missing winter data. Thus, Figure 2 starts at 1995. Although hourly temperature data have been collected from moored buoys since 1980 in most of the Great Lakes, these data contain wide gaps for a variety of reasons, including scheduled maintenance, sensor malfunctions, and natural elements (e.g., winter conditions). These data gaps prevent reliable and consistent annual averages from being calculated from buoy data.

3. Since the first water level gauges were installed in 1860, several major engineering projects have been undertaken to modify the Great Lakes basin for use by cities and residents in the area. The most prominent of these have been the dredging efforts in the St. Clair River, which connects Lakes Michigan and Huron to Lake St. Clair, to support commercial navigation. At least some of the decrease in water levels in Lake Michigan and Lake Huron has been attributed to this dredging. Specifically, the St. Clair river opening was enlarged in the 1910s, 1930s, and 1960s, contributing to greater outflows from Lakes Michigan and Huron (Quinn, 1985). Similar projects have also occurred in other areas of the Great Lakes basin, although they have not been linked directly to changes in lake water levels.

4. In addition to changes in channel depth, recent studies have found that dredging projects significantly increased the erosion in channel bottoms. The combination of dredging and erosion is estimated to have resulted in a 20-inch decrease in water levels for Lakes Michigan and Huron between 1908 and 2012 (Egan, 2013).

10. Sources of Uncertainty

Individual water level sensors are estimated to be relatively accurate. The gauges have an estimated accuracy of ±0.006 meters for individual measurements, which are conducted every six minutes, and ±0.003 meters for calculated monthly means (NOAA, 2013). In the instance of sensor or other equipment failure, NOAA does not interpolate values to fill in data gaps. However, because data gaps are at a small temporal resolution (minutes to hours), they have little effect on indicator values, which have a temporal resolution of months to years.

Surface water temperature observations from satellites are subject to navigation, timing, and calibration errors. An automated georeferencing process was used to reduce navigation errors to 2.6 kilometers. When compared with buoy data, for reference, satellite data from the pixel nearest the buoy location differ by less than 0.5°C. The root mean square difference ranges from 1.10 to 1.76°C with correlation coefficients above 0.95 for all buoys (Schwab et al., 1999).

11. Sources of Variability

Water levels are sensitive to changes in climate, notably temperature (affecting evaporation and ice cover) and precipitation. Natural variation in climate of the Great Lakes basin will affect recorded water levels. In addition to climate, water levels are also affected by changing hydrology, including dredging of channels between lakes, the reversal of the Chicago River, changing land-use patterns, and industrial water usage. However, the long time span of this indicator allows for an analysis of trends over more than a century. Water withdrawals could also influence water levels, but arguably not as large a role as climate or dredging because nearly all (95 percent) of the water withdrawn from the Great Lakes is returned via discharge or runoff (U.S. EPA, 2009).

Surface water temperature is sensitive to many natural environmental factors, including precipitation and water movement. Natural variations in climate of the Great Lakes basin will affect recorded water temperature. In addition to climate, water temperature is also affected by human water use. For example, industries have outflows into several of the Great Lakes, which may affect water temperatures in specific areas of the lake.

12. Statistical/Trend Analysis

Water Levels

Multivariate adaptive regression splines (MARS) (Friedman, 1991; Milborrow, 2012) were used within each lake to model non-linear behavior in water levels through time. The MARS regression technique was used because of its ability to partition the data into separate regions that can be treated independently. MARS regressions were used to identify when the recent period of declining water levels began. For three of the four Great Lakes basins (Michigan-Huron, Erie, and Superior), 1986 marked the beginning of a distinct, statistically significant negative trend in water levels. The MARS analysis suggests that water levels in Lake Ontario have remained relatively constant since the late 1940s.

To characterize the extent to which recent water levels represent deviation from long-term mean values, EPA used t-tests to compare recent average water levels (2004–2013) against long-term averages (1860–2013). Statistically significant differences (p < 0.01) were found in the annual data for Lakes Superior and Michigan-Huron.

Surface Water Temperatures

Table TD-2 below shows the slope, p-value, and total change from an ordinary least-squares linear regression of each lake's annual data. Trends for all lakes except Erie are statistically significant to a 95 percent level; Erie is significant to a 90 percent level.

Table TD-2. Linear Regression of Annual Satellite Data, 1992–2013

Lake	Slope (°F/year)	P-value	Total change (°F) (slope x 21 years)
Erie	0.070	0.068	1.551
Huron	0.147	0.0015	3.242
Michigan	0.125	0.021	2.750
Ontario	0.175	0.0002	3.860
Superior	0. 192	0.0011	4.228

For the daily temperature graphs on the right side of Figure 2, paired t-tests were used to compare recent average daily surface water temperature (2005–2013) against the previous 10 years (1995–2004). All five lakes showed highly significant differences between the two time periods ($p < 0.0001$).

References

Egan, D. 2013. Does Lake Michigan's record low mark beginning of new era for Great Lakes? Milwaukee Journal Sentinel. July 27, 2013.

Friedman, J.H. 1991. Multivariate adaptive regression splines. Ann. Stat. 19(1):1–67.

Gronewold, A.D., V. Fortin, B. Lofgren, A. Clites, C.A. Stow, and F. Quinn. 2013. Coasts, water levels, and climate change: A Great Lakes perspective. Climatic Change 120:697–711.

Milborrow, S. 2012. Earth: Multivariate adaptive regression spline models. Derived from mda:mars by Trevor Hastie and Rob Tibshirani. R package version 3.2-3. http://CRAN.R-project.org/package=earth.

NOAA (National Oceanic and Atmospheric Administration). 2013. Environmental measurement systems: Sensor specifications and measurement algorithms. National Ocean Service, Center for Operational Oceanographic Products and Services. http://tidesandcurrents.noaa.gov/publications/CO-OPS_Measure_Spec_07_July_2013.pdf.

Quinn, F.H. 1985. Temporal effects of St. Clair River dredging on Lakes St. Clair and Erie water levels and connecting channel flow. J. Great Lakes Res. 11(3):400–403.

Schwab, D.J., G.A. Leshkevich, and G.C. Muhr. 1999. Automated mapping of surface water temperature in the Great Lakes. J. Great Lakes Res. 25(3):468–481.

U.S. EPA. 2009. State of the Great Lakes 2009: Water withdrawals. www.epa.gov/greatlakes/solec/sogl2009/7056waterwithdrawals.pdf.

Bird Wintering Ranges

Identification

1. Indicator Description

This indicator examines changes in the winter ranges of North American birds from the winter of 1966–1967 to 2005. Changes in climate can affect ecosystems by influencing animal behavior and ranges. Birds are a particularly strong indicator of environmental change for several reasons described in the indicator text. This indicator focuses in particular on latitude—how far north or south birds travel—and distance from the coast. Inland areas tend to experience more extreme cold than coastal areas, but birds may shift inland over time as winter temperature extremes grow less severe.

Components of this indicator include:

- Shifts in the latitude of winter ranges of North American birds over the past half-century (Figure 1).
- Shifts in the distance to the coast of winter ranges of North American birds over the past half-century (Figure 2).

2. Revision History

April 2010:   Indicator posted.
May 2014:   Updated with data through 2013.

Data Sources

3. Data Sources

This indicator is based on data collected by the annual Christmas Bird Count (CBC), managed by the National Audubon Society. Data used in this indicator are collected by citizen scientists who systematically survey certain areas and identify and count widespread bird species. The CBC has been in operation since 1900, but data used in this indicator begin in winter 1966–1967.

4. Data Availability

Complete CBC data are available in both print and electronic formats. Historical CBC data have been published in several periodicals—*Audubon Field Notes*, *American Birds*, and *Field Notes*—beginning in 1998. Additionally, historical, current year, and annual summary CBC data are available online at: http://birds.audubon.org/christmas-bird-count. Descriptions of data are available with the data queried online. The appendix to National Audubon Society (2009) provides 40-year trends for each species, but not the full set of data by year. EPA obtained the complete data set for this indicator, with trends and species-specific data through 2013, directly from the National Audubon Society.

A similar analysis is available from an interagency consortium at: www.stateofthebirds.org/2010.

Methodology

5. Data Collection

This indicator is based on data collected by the annual CBC, managed by the National Audubon Society. Data used in this indicator are collected by citizen scientists who systematically survey certain areas and identify and count widespread bird species. Although the indicator relies on human observation rather than precise measuring instruments, the people who collect the data are skilled observers who follow strict protocols that are consistent across time and space. These data have supported many peer-reviewed studies, a list of which can be found on the National Audubon Society's website at: http://birds.audubon.org/christmas-bird-count-bibliography-scientific-articles.

Bird surveys take place each year in approximately 2,000 different locations throughout the contiguous 48 states and the southern portions of Alaska and Canada. All local counts take place between December 14 and January 5 of each winter. Each local count takes place over a 24-hour period in a defined "count circle" that is 15 miles in diameter. A variable number of volunteer observers separate into field parties, which survey different areas of the count circle and tally the total number of individuals of each species observed (National Audubon Society, 2009). This indicator covers 305 bird species, which are listed in Appendix 1 of National Audubon Society (2009). These species were included because they are widespread and they met specific criteria for data availability.

The entire study description, including a list of species and a description of sampling methods and analyses performed, can be found in National Audubon Society (2009) and references therein. Information on this study is also available on the National Audubon Society website at: http://birdsandclimate.audubon.org/index.html. For additional information on CBC survey design and methods, see the reports classified as "Methods" in the list at: http://birds.audubon.org/christmas-bird-count-bibliography-scientific-articles.

6. Indicator Derivation

At the end of the 24-hour observation period, each count circle tallies the total number of individuals of each species seen in the count circle. Audubon scientists then run the data through several levels of analysis and quality control to determine final count numbers from each circle and each region. Data processing steps include corrections for different levels of sampling effort—for example, if some count circles had more observers and more person-hours of effort than others. Population trends over the 40-year period of this indicator and annual indices of abundance were estimated for the entire survey area with hierarchical models in a Bayesian analysis using Markov chain Monte Carlo techniques (National Audubon Society, 2009).

This indicator is based on the center of abundance for each species, which is the center of the population distribution at any point in time. In terms of latitude, half of the individuals in the population live north of the center of abundance and the other half live to the south. Similarly, in terms of longitude, half of the individuals live west of the center of abundance, and the other half live to the east. The center of abundance is a common way to characterize the general location of a population. For example, if a population were to shift generally northward, the center of abundance would be expected to shift northward as well.

This indicator examines the center of abundance from two perspectives:

- Latitude—testing the hypothesis that bird populations are moving northward along with the observed rise in overall temperatures throughout North America.

- Distance from coast—testing the hypothesis that bird populations are able to move further from the coast as a generally warming climate moderates the inland temperature extremes that would normally occur in the winter.

This indicator reports the position of the center of abundance for each year, relative to the position of the center of abundance in 1966 (winter 1966–1967). The change in position is averaged across all 305 species for changes in latitude (Figure 1) and across 272 species for changes in distance from the coast (Figure 2). The indicator excludes 33 species from the analysis of distance from the coast because these species depend on a saltwater or brackish water habitat. Lake shorelines (including the Great Lakes) were not considered coastlines for the purposes of the "distance from coast" metric.

Figures 1 and 2 show average distances moved north and moved inland, based on an unweighted average of all species. Thus, no adjustments are made for population differences across species.

No attempt was made to generate estimates outside the surveyed area. The indicator does not include Mexico or northern parts of Alaska and Canada because data for these areas were too sparse to support meaningful trend analysis. Due to its distance from the North American continent, Hawaii is also omitted from the analysis. No attempt was made to estimate trends prior to 1966 (i.e., prior to the availability of complete spatial coverage and standardized methods), and no attempt was made to project trends into the future.

The entire study description, including analyses performed, can be found in National Audubon Society (2009) and references therein. Information on this study is also available on the National Audubon Society website at: http://birdsandclimate.audubon.org/index.html.

7. Quality Assurance and Quality Control

As part of the overall data compilation effort, Audubon scientists have performed several statistical analyses to ensure that potential error and variability are adequately addressed. Quality assurance/quality control procedures are described in National Audubon Society (2009) and in a variety of methodology reports listed at: http://birds.audubon.org/christmas-bird-count-bibliography-scientific-articles.

Analysis

8. Comparability Over Time and Space

The CBC has been in operation since 1900, but data used in this indicator begin in winter 1966–1967. The National Audubon Society chose this start date to ensure sufficient sample size throughout the survey area as well as consistent methods, as the CBC design and methodology have remained generally consistent since the 1960s. All local counts take place between December 14 and January 5 of each winter, and they follow consistent methods regardless of the location.

9. Data Limitations

Factors that may impact the confidence, application, or conclusions drawn from this indicator are as follows:

1. Many factors can influence bird ranges, including food availability, habitat alteration, and interactions with other species. Some of the birds covered in this indicator might have moved northward or inland for reasons other than changing temperatures.

2. This indicator does not show how responses to climate change vary among different types of birds. For example, National Audubon Society (2009) found large differences between coastal birds, grassland birds, and birds adapted to feeders, which all have varying abilities to adapt to temperature changes. This Audubon report also shows the large differences between individual species—some of which moved hundreds of miles while others did not move significantly at all.

3. Some data variations are caused by differences between count circles, such as inconsistent level of effort by volunteer observers, but these differences are carefully corrected in Audubon's statistical analysis.

4. While observers attempt to identify and count every bird observed during the 24-hour observation period, rare and nocturnal species may be undersampled. Gregarious species (i.e., species that tend to gather in large groups) can also be difficult to count, and they could be either overcounted or undercounted, depending on group size and the visibility of their roosts. These species tend to congregate in known and expected locations along CBC routes, however, so observers virtually always know to check these spots. Locations with large roosts are often assigned to observers with specific experience in estimating large numbers of birds.

5. The tendency for saltwater-dependent species to stay near coastlines could impact the change in latitude calculation for species living near the Gulf of Mexico. By integrating these species into the latitudinal calculation, Figure 1 may understate the total extent of northward movement of species.

10. Sources of Uncertainty

The sources of uncertainty in this indicator have been analyzed, quantified, and accounted for to the extent possible. The statistical significance of the trends suggests that the conclusions one might draw from this indicator are robust.

One potential source of uncertainty in these data is uneven effort among count circles. Various studies that discuss the best ways to account for this source of error have been published in peer-reviewed journals. Link and Sauer (1999) describe the methods that Audubon used to account for variability in effort.

11. Sources of Variability

Rare or difficult-to-observe bird species could lead to increased variability. For this analysis, the National Audubon Society included only 305 widespread bird species that met criteria for abundance and the availability of data to enable the detection of meaningful trends.

12. Statistical/Trend Analysis

Appendix 1 of National Audubon Society (2009) documents the statistical significance of trends in the wintering range for each species included in an earlier version of this indicator. Using annual data points for each species, EPA applied an ordinary least-squares regression to determine the statistical significance of each species' movement, as well as the statistical significance of each overall trend. Tables TD-1 and TD-2 present these two analyses. Both of these tables are based on an analysis of all 305 species that the National Audubon Society studied.

Table TD-1. Statistical Analyses of Aggregate (All Species) Trends

Indicator component	Regression slope	P-value	Total miles moved
Northward (latitude)	0.993 miles/year	<0.0001	46.7
Inward from the coast	0.231 miles/year	<0.0001	10.9

Table TD-2. Statistical Analyses of Species-Specific Trends

Statistical calculation	Figure 1	Figure 2
Species with significant* northward/inward movement	186	174
Species with significant* southward/coastward movement	82	97
Species with northward/inward movement >200 miles	48	3

*In Tables TD-1 and TD2, "significant" refers to 95 percent confidence (p < 0.05).

The shaded bands in Figures 1 and 2 show 95 percent upper and lower credible intervals, which are Bayesian statistical outputs that are analogous to 95 percent confidence intervals.

References

Link, W.A., and J.R. Sauer. 1999. Controlling for varying effort in count surveys: An analysis of Christmas Bird Count data. J. Agric. Biol. Envir. S. 4:116–125.

National Audubon Society. 2009. Northward shifts in the abundance of North American birds in early winter: a response to warmer winter temperatures? www.audubon.org/bird/bacc/techreport.html.

Leaf and Bloom Dates

Identification

1. Indicator Description

This indicator examines the timing of first leaf dates and flower bloom dates in lilacs and honeysuckle plants in the contiguous 48 states between 1900 and 2013. The first leaf date in these plants relates to the timing of events that occur in early spring, while the first bloom date is consistent with the timing of later spring events, such as the start of growth in forest vegetation. Lilacs and honeysuckles are especially useful as indicators of spring events because they are widely distributed across most of the contiguous 48 states and widely studied in the peer-reviewed literature. Scientists have very high confidence that recent warming trends in global climate have contributed to the earlier arrival of spring events (IPCC, 2014).

Components of this indicator include:

- Trends in first leaf dates and first bloom dates since 1900, aggregated across the contiguous 48 states (Figure 1).
- A map showing changes in first leaf dates between 1951–1960 and 2004–2013 (Figure 2).
- A map showing changes in first bloom dates between 1951–1960 and 2004–2013 (Figure 3).

2. Revision History

April 2010:   Indicator posted.
December 2011:   Updated with data through 2010.
December 2013:   Original Figures 1 and 2 (leaf and bloom date time series) combined and
                       updated with data through 2013; new maps (Figures 2 and 3) added.

Data Sources

3. Data Sources

This indicator is based on leaf and bloom observations that were compiled by the USA National Phenology Network (USA-NPN) and climate data that were provided by the U.S. Historical Climatology Network (USHCN) and other databases maintained by the National Oceanic and Atmospheric Administration's (NOAA's) National Climatic Data Center (NCDC). Data for this indicator were analyzed using a method described by Schwartz et al. (2013).

4. Data Availability

Phenological Observations

This indicator is based in part on observations of lilac and honeysuckle leaf and bloom dates, to the extent that these observations contributed to the development of models. USA-NPN provides online access to historical phenological observations at: www.usanpn.org/data.

Temperature Data

This indicator is based in part on historical daily temperature records, which are publicly available online through NCDC. For example, USHCN data are available online at: www.ncdc.noaa.gov/oa/climate/research/ushcn, with no confidentiality issues limiting accessibility. Appropriate metadata and "readme" files are appended to the data so that they are discernible for analysis. For example, see: ftp://ftp.ncdc.noaa.gov/pub/data/ushcn/v2/monthly/readme.txt. Summary data from other sets of weather stations can be obtained from NCDC at: www.ncdc.noaa.gov.

Model Results

The processed leaf and bloom date data set is not publicly available. EPA obtained the model outputs by contacting Dr. Mark Schwartz at the University of Wisconsin–Milwaukee, who developed the analysis and created the original time series and maps. Results of this analysis have been published in Schwartz et al. (2013) and other papers.

Methodology

5. Data Collection

This indicator was developed using models that relate phenological observations (leaf and bloom dates) to weather and climate variables. These models were developed by analyzing the relationships between two types of measurements: 1) observations of the first leaf emergence and the first flower bloom of the season in lilacs and honeysuckles and 2) temperature data. The models were developed using measurements collected throughout the portions of the Northern Hemisphere where lilacs and/or honeysuckles grow, then applied to temperature records from a larger set of stations throughout the contiguous 48 states.

Phenological Observations

First leaf date is defined as the date on which leaves first start to grow beyond their winter bud tips. First bloom date is defined as the date on which flowers start to open. Ground observations of leaf and bloom dates were gathered by government agencies, field stations, educational institutions, and trained citizen scientists; these observations were then compiled by organizations such as the USA-NPN. These types of phenological observations have a long history and have been used to support a wide range of peer-reviewed studies. See Schwartz et al. (2013) and references cited therein for more information about phenological data collection methods.

Temperature Data

Weather data used to construct, validate, and then apply the models—specifically daily maximum and minimum temperatures—were collected from officially recognized weather stations using standard meteorological instruments. These data have been compiled by NCDC databases such as the USHCN and TD3200 Daily Summary of the Day data from other cooperative weather stations. As described in the methods for an earlier version of this analysis (Schwartz et al., 2006), station data were used rather than gridded values, "primarily because of the undesirable homogenizing effect that widely available coarse-resolution grid point data can have on spatial differences, resulting in artificial uniformity of processed outputs..." (Schwartz and Reiter, 2000; Schwartz and Chen, 2002; Menzel et al., 2003). Ultimately, 799 weather stations were selected according to the following criteria:

- Provide for the best temporal and spatial coverage possible. At some stations, the period of record includes most of the 20th century.

- Have at least 25 of 30 years during the 1981–2010 baseline period, with no 30-day periods missing more than 10 days of data.

- Have sufficient spring–summer warmth to generate valid model output.

For more information on the procedures used to obtain temperature data, see Schwartz et al. (2013) and references cited therein.

6.   Indicator Derivation

Daily temperature data and observations of first leaf and bloom dates were used to construct and validate a set of models that relate phenological observations to weather and climate variables (specifically daily maximum and minimum temperatures). These models were developed for the entire Northern Hemisphere and validated at 378 sites in Germany, Estonia, China, and the United States.

Once the models were validated, they were applied to locations throughout the contiguous 48 states using temperature records from 1900 to 2013. Even if actual phenological observations were not collected at a particular station, the models essentially predict phenological behavior based on observed daily maximum and minimum temperatures, allowing the user to estimate the date of first leaf and first bloom for each year at that location. The value of these models is that they can estimate the onset of spring events in locations and time periods where actual lilac and honeysuckle observations are sparse. In the case of this indicator, the models have been applied to a time period that is much longer than most phenological observation records. The models have also been extended to areas of the contiguous 48 states where lilacs and honeysuckles do not actually grow—mainly parts of the South and the West coast where winter is too warm to provide the extended chilling that these plants need in order to bloom the following spring. This step was taken to provide more complete spatial coverage.

This indicator was developed by applying phenological models to several hundred sites in the contiguous 48 states where sufficient weather data have been collected. The exact number of sites varies from year to year depending on data availability (the minimum was 297 sites in 1901; the maximum was 771 sites in 1991).

After running the models, analysts looked at each location and compared the first leaf date and first bloom date in each year with the average leaf date and bloom date for 1981 to 2010, which was established as a "climate normal" or baseline. This step resulted in a data set that lists each station along with the "departure from normal" for each year—measured in days—for each component of the indicator (leaf date and bloom date). Note that 1981 to 2010 represents an arbitrary baseline for comparison, and choosing a different baseline period would shift the observed long-term trends up or down but would not alter the shape, magnitude, or statistical significance of the trends.

Figure 1. First Leaf and Bloom Dates in the Contiguous 48 States, 1900–2013

EPA obtained a data set listing annual departure from normal for each station, then performed some additional steps to create Figure 1. For each component of the indicator (leaf date and bloom date), EPA aggregated the data for each year to determine an average departure from normal across all stations. This step involved calculating an unweighted arithmetic mean of all stations with data in a given year. The aggregated annual trend line appears as a thin curve in each figure. To smooth out some of the year-to-year variability, EPA also calculated a nine-year weighted moving average for each component of the indicator. This curve appears as a thick line in each figure, with each value plotted at the center of the corresponding nine-year window. For example, the average from 2000 to 2008 is plotted at year 2004. This nine-year average was constructed using a normal curve weighting procedure that preferentially weights values closer to the center of the window. Weighting coefficients for values 1 through 9, respectively, were as follows: 0.0076, 0.036, 0.1094, 0.214, 0.266, 0.214, 0.1094, 0.036, 0.0076. This procedure was recommended by the authors of Schwartz et al. (2013) as an appropriate way to reduce some of the "noise" inherent in annual phenology data.

EPA used endpoint padding to extend the nine-year smoothed lines all the way to the ends of the period of record. Per the data provider's recommendation, EPA calculated smoothed values centered at 2010, 2011, 2012, and 2013 by inserting the 2009–2013 average into the equation in place of the as-yet unreported annual data points for 2014 and beyond. EPA used an equivalent approach at the beginning of the time series.

Figures 2 and 3. Change in First Leaf and Bloom Dates Between 1951–1960 and 2004–2013

To show spatial patterns in leaf and bloom changes, Figures 2 and 3 compare the most recent decade of data with the decade from 1951 to 1960 at individual stations. The 1950s were chosen as a baseline period to be consistent with the analysis published by Schwartz et al. (2013), who noted that broad changes in the timing of spring events appeared to start around the 1950s. To create the maps, EPA calculated the average departure from normal during each 10-year period and then calculated the difference between the two periods. The maps are restricted to stations that had at least eight years of valid data in both 10-year periods; 561 stations met these criteria.

For more information on the procedures used to develop, test, and apply the models for this indicator, see Schwartz et al. (2013) and references cited therein.

Indicator Development

The 2010 edition of EPA's *Climate Change Indicators in the United States* report presented an earlier version of this indicator based on an analysis published in Schwartz et al. (2006). That analysis was referred to as the Spring Indices (SI). The team that developed the original SI subsequently developed an enhanced version of their algorithm, which is referred to as the Extended Spring Indices (SI-x). EPA adopted the SI-x approach for the 2012 edition of *Climate Change Indicators in the United States*. The SI-x represents an extension of the original SI because it can now characterize the timing of spring events in areas where lilacs and honeysuckles do not grow. Additional details about the SI-x are discussed in Schwartz et al. (2013).

For the 2014 edition of this indicator, EPA added a set of maps (Figures 2 and 3) to provide a more robust depiction of regional variations. These maps were published in Schwartz et al. (2013) and have since been updated with more recent data.

7. Quality Assurance and Quality Control

Phenological Observations

Quality assurance and quality control (QA/QC) procedures for phenological observations are not readily available.

Temperature Data

Most of the daily maximum and minimum temperature values were evaluated and cleaned to remove questionable values as part of their source development. For example, several papers have been written about the methods of processing and correcting historical climate data for the USHCN. NCDC's website (www.ncdc.noaa.gov/oa/climate/research/ushcn) describes the underlying methodology and cites peer-reviewed publications justifying this approach.

Before applying the model, all temperature data were checked to ensure that no daily minimum temperature value was larger than the corresponding daily maximum temperature value (Schwartz et al., 2006).

Model Results

QA/QC procedures are not readily available regarding the use of the models and processing the results. These models and results have been published in numerous peer-reviewed studies, however, suggesting a high level of QA/QC and review. For more information about the development and application of these models, see Schwartz et al. (2013), McCabe et al. (2012), and the references cited therein.

Analysis

8. Comparability Over Time and Space

Phenological Observations

For consistency, the phenological observations used to develop this indicator were restricted to certain cloned species of lilac and honeysuckle. Using cloned species minimizes the influence of genetic differences in plant response to temperature cues, and it helps to ensure consistency over time and space.

Temperature Data

The USHCN has undergone extensive testing to identify errors and biases in the data and either remove these stations from the time series or apply scientifically appropriate correction factors to improve the utility of the data. In particular, these corrections address changes in the time-of-day of observation, advances in instrumentation, and station location changes.

Homogeneity testing and data correction methods are described in more than a dozen peer-reviewed scientific papers by NCDC. Data corrections were developed to specifically address potential problems in trend estimation of the rates of warming or cooling in the USHCN. Balling and Idso (2002) compared the USHCN data with several surface and upper-air data sets and showed that the effects of the various USHCN adjustments produce a significantly more positive, and likely spurious, trend in the USHCN data. In contrast, a subsequent analysis by Vose et al. (2003) found that USHCN station history information is reasonably complete and that the bias adjustment models have low residual errors.

Further analysis by Menne et al. (2009) suggests that:

> ...the collective impact of changes in observation practice at USHCN stations is systematic and of the same order of magnitude as the background climate signal. For this reason, bias adjustments are essential to reducing the uncertainty in U.S. climate trends. The largest biases in the HCN are shown to be associated with changes to the time of observation and with the widespread changeover from liquid-in-glass thermometers to the maximum minimum temperature sensor (MMTS). With respect to [USHCN] Version 1, Version 2 trends in maximum temperatures are similar while minimum temperature trends are somewhat smaller because of an apparent overcorrection in Version 1 for the MMTS instrument change, and because of the systematic impact of undocumented station changes, which were not addressed [in] Version 1.

USHCN Version 2 represents an improvement in this regard.

Some observers have expressed concerns about other aspects of station location and technology. For example, Watts (2009) expresses concern that many U.S. weather stations are sited near artificial heat sources such as buildings and paved areas, potentially biasing temperature trends over time. In response to these concerns, NOAA analyzed trends for a subset of stations that Watts had determined to be "good or best," and found the temperature trend over time to be very similar to the trend across the full set of USHCN stations (www.ncdc.noaa.gov/oa/about/response-v2.pdf). While it is true that many stations are not optimally located, NOAA's findings support the results of an earlier analysis by Peterson (2006) that found no significant bias in long-term trends associated with station siting once NOAA's homogeneity adjustments have been applied.

Model Results

The same model was applied consistently over time and space. Figure 1 generalizes results over space by averaging station-level departures from normal in order to determine the aggregate departure from normal for each year. This step uses a simple unweighted arithmetic average, which is appropriate given the national scale of this indicator and the large number of weather stations spread across the contiguous 48 states.

9. Data Limitations

Factors that may impact the confidence, application, or conclusions drawn from this indicator are as follows:

1. Plant phenological events are studied using several data collection methods, including satellite images, models, and direct observations. The use of varying data collection methods in addition to the use of different phenological indicators (such as leaf or bloom dates for different types of plants) can lead to a range of estimates of the arrival of spring.

2. Climate is not the only factor that can affect phenology. Observed variations can also reflect plant genetics, changes in the surrounding ecosystem, and other factors. This indicator minimizes genetic influences by relying on cloned plant species, however (that is, plants with no genetic differences).

10. Sources of Uncertainty

Error estimates are not readily available for the underlying temperature data upon which this indicator is based. It is generally understood that uncertainties in the temperature data increase as one goes back in time, as there are fewer stations early in the record. However, these uncertainties are not sufficient to mislead the user about fundamental trends in the data.

In aggregating station-level "departure from normal" data into an average departure for each year, EPA calculated the standard error of each component of Figure 1 (leaf date and bloom date) in each year. For both components, standard errors range from 0.3 days to 0.5 days, depending on the year.

Uncertainty has not been calculated for the individual station-level changes shown in Figures 2 and 3.

Schwartz et al. (2013) provide error estimates for the models. The use of modeled data should not detract from the conclusions that can be inferred from the indicator. These models have been extensively tested and refined over time and space such that they offer good certainty.

11. Sources of Variability

Temperatures naturally vary from year to year, which can strongly influence leaf and bloom dates. To smooth out some of the year-to-year variability, EPA calculated a nine-year weighted moving average for each component of this indicator in Figure 1, and EPA created the maps in Figures 2 and 3 based on 10-year averages for each station.

12. Statistical/Trend Analysis

Statistical testing of individual station trends within the contiguous 48 states suggests that many of these trends are not significant. Other studies (e.g., Schwartz et al., 2006) have come to similar conclusions, finding that trends in the earlier onset of spring at individual stations are much stronger in Canada and parts of Eurasia than they are in the contiguous 48 states. In part as a result of these findings, Figure 1 focuses on aggregate trends across the contiguous 48 states, which should be more statistically robust than individual station trends. However, the aggregate trends still are not statistically significant ($p < 0.05$) over the entire period of record, based on a simple t-test.

References

Balling, Jr., R.C., and C.D. Idso. 2002. Analysis of adjustments to the United States Historical Climatology Network (USHCN) temperature database. Geophys. Res. Lett. 29(10):1387.

IPCC (Intergovernmental Panel on Climate Change). 2014. Climate change 2014: Impacts, adaptation, and vulnerability. Working Group II contribution to the IPCC Fifth Assessment Report. Cambridge, United Kingdom: Cambridge University Press. www.ipcc.ch/report/ar5/wg2.

McCabe, G.J., T.R. Ault, B.I. Cook, J.L. Betancourt, and M.D. Schwartz. 2012. Influences of the El Niño Southern Oscillation and the Pacific Decadal Oscillation on the timing of the North American spring. Int. J. Climatol. 32:2301–2310.

Menne, M.J., C.N. Williams, Jr., and R.S. Vose. 2009. The U.S. Historical Climatology Network monthly temperature data, version 2. B. Am. Meteorol. Soc. 90:993-1107. ftp://ftp.ncdc.noaa.gov/pub/data/ushcn/v2/monthly/menne-etal2009.pdf.

Menzel, A., F. Jakobi, R. Ahas, et al. 2003. Variations of the climatological growing season (1951–2000) in Germany compared to other countries. Int. J. Climatol. 23:793–812.

Peterson, T.C. 2006. Examination of potential biases in air temperature caused by poor station locations. B. Am. Meteorol. Soc. 87:1073–1080. http://journals.ametsoc.org/doi/pdf/10.1175/BAMS-87-8-1073.

Schwartz, M.D., and X. Chen. 2002. Examining the onset of spring in China. Clim. Res. 21:157–164.

Schwartz, M.D., and B.E. Reiter. 2000. Changes in North American spring. Int. J. Climatol. 20:929–932.

Schwartz, M.D., R. Ahas, and A. Aasa. 2006. Onset of spring starting earlier across the Northern Hemisphere. Glob. Change Biol. 12:343–351.

Schwartz, M.D., T.R. Ault, and J.L. Betancourt. 2013. Spring onset variations and trends in the continental United States: Past and regional assessment using temperature-based indices. Int. J. Climatol. 33:2917–2922.

Vose, R.S., C.N. Williams, Jr., T.C. Peterson, T.R. Karl, and D.R. Easterling. 2003. An evaluation of the time of observation bias adjustment in the U.S. Historical Climatology Network. Geophys. Res. Lett. 30(20):2046.

Watts, A. 2009. Is the U.S. surface temperature record reliable? The Heartland Institute. http://wattsupwiththat.files.wordpress.com/2009/05/surfacestationsreport_spring09.pdf.

Cherry Blossom Bloom Dates in Washington, D.C.

Identification

1. Description

This regional feature highlights the peak bloom date (PBD) for the most common species of cherry tree planted around the Tidal Basin in Washington, D.C., from 1921 to 2014. The PBD provides insight into how shifting climate patterns are affecting the timing of cherry blossom blooming in one particular community as an example of an event associated with the onset of spring. Shifts in phenological events such as bloom dates can have important implications for ecosystem processes and could have economic and cultural consequences. For reference, this feature also shows the start and end dates of the National Cherry Blossom Festival, which is planned to coincide with the predicted PBD each year.

2. Revision History

December 2013:   Feature proposed.
April 2014:   Updated with 2014 data.

Data Sources

3. Data Sources

Data were provided by the National Park Service (NPS) within the U.S. Department of the Interior (DOI), which cares for the cherry trees around Washington's Tidal Basin. The NPS has continuously monitored PBD since 1921 for the cherry trees around the Tidal Basin.

The NPS also records the dates for the National Cherry Blossom Festival, with data spanning 1934–2014. There was a five-year gap from 1942 to 1946 when the festival was canceled due to World War II.

4. Data Availability

All cherry blossom PBD data, as well as National Cherry Blossom Festival dates, are maintained by the NPS and can be downloaded from the Web at: www.nps.gov/cherry/cherry-blossom-bloom.htm. All data are publicly available. Festival dates for 2012–2014 had not been added to the online record at the time this regional feature was developed, but were provided by the organizers of the festival (contact information at: www.nationalcherryblossomfestival.org).

Methodology

5. Data Collection

NPS horticulturalists carefully monitor the 3,770 cherry trees around the Tidal Basin. The most prevalent species—and the one covered by this feature—is Yoshino (*Prunus x yedoensis*), which constitutes 70.6

percent of Washington's cherry trees. NPS staff have also monitored another species, Kwanzan (*Prunus serrulata 'Kwanzan'*), representing 12.8 percent of the trees present, but the Kwanzan data are missing several years (including all years since 2012), so they were not included in this regional feature.

NPS horticulturalists examine a specific set of Yoshino trees daily and evaluate them with respect to five stages of cherry blossom development: green buds, florets visible, extension of florets, peduncle elongation, and puffy white. They use this approach to determine the official PBD, which is defined as the day when 70 percent of the Yoshino cherry tree blossoms are open in full bloom. A pictorial description of the phases of cherry blossom development, as well as other general information about blooming periods, is available at: www.nps.gov/cherry/cherry-blossom-bloom.htm.

6. Derivation

Figure 1 plots the annual PBD for the Yoshino trees from 1921 to 2014, along with the annual start and end dates of the National Cherry Blossom Festival.

For consistency, EPA converted bloom and festival dates into Julian days to support graphing and calculations. By this method, January 1 = day 1, etc. The method also accounts for leap years, such that March 31 = day 90 in a non-leap year and day 91 in a leap year, for example. Figure 1 actually plots Julian dates, but the corresponding non-leap year calendar dates have been added to the y-axis to provide a more familiar frame of reference. This means that a PBD of March 31 in a leap year will actually be plotted at the same level as April 1 from a non-leap year, for example, and it will appear to be plotted at April 1 with respect to the y-axis.

7. Quality Assurance and Quality Control

By monitoring the five different stages of cherry blossom bud development, NPS horticulturalists are able to forecast, and ultimately pinpoint, PBD with minimal uncertainty.

Analysis

8. Comparability Over Time and Space

The NPS has recorded PBD annually for Yoshino cherry trees since 1921, using a consistent definition of PBD, examining the same group of Yoshino cherry trees, and using the same set of bud break criteria throughout the period of record. These consistent practices allow for comparability over time.

Start and end dates for the National Cherry Blossom Festival have been provided for reference only. While these dates add an interesting cultural and economic element to this regional feature, they fundamentally reflect human decisions based on economic and social factors that have changed over time. In particular, the festival has evolved from a single day or weekend to a multi-week event.

This regional feature is limited to a small geographic area. Methods have been applied consistently within this area.

9. Data Limitations

Factors that may impact the confidence, application, or conclusions drawn from this feature are as follows:

1. The timing of PBD for cherry trees can be affected by a variety of weather and climate factors. This feature does not necessarily pinpoint a single main cause of the observed trends, although winter and early spring temperatures are believed to play a key role.

2. The PBD does not provide information on non-climate factors that may affect cherry tree phenology (the timing of key developmental events) or health.

10. Sources of Uncertainty

Because PBD is clearly defined, and NPS horticulturalists have used a single, consistent definition over time, there is little uncertainty in either the definition or identification of PBD. Uncertainty in the measurements has not been explicitly quantified, however.

11. Sources of Variability

Because PBD is highly sensitive to changes in temperature, natural variations in seasonal temperatures contribute to year-to-year variations in PBD. Although the PBD for these cherry trees is primarily affected by temperature, other aspects of changing climate could also affect the annual blooming date. Extended growing periods or warmer autumns may indirectly affect PBD by altering other stages of cherry tree development (Chung et al., 2011).

12. Statistical/Trend Analysis

EPA calculated the long-term trend in PBD for Yoshino cherry trees by ordinary least-squares linear regression to support a statement in the "Key Points" text. The 1921–2014 trend had a slope of -0.053 days/year, with $p = 0.051$. Thus, the trend is significant to a 90 percent level but not to a 95 percent level.

References

Chung, U., L. Mack, J.I. Yun, and S. Kim. 2011. Predicting the timing of cherry blossoms in Washington, D.C. and Mid-Atlantic states in response to climate change. PLOS ONE 6(11):e27439.

www.ingramcontent.com/pod-product-compliance
Lightning Source LLC
Chambersburg PA
CBHW080635180526
45168CB00008B/3177